HIDDEN
SECRETS

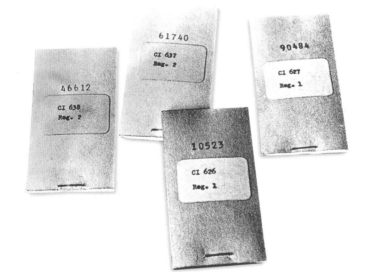

pst!
Feind hört mit

HIDDEN SECRETS

DAVID OWEN

FIREFLY BOOKS

A FIREFLY BOOK

Published by Firefly Books Ltd, 2002

Copyright © 2002 Quintet Publishing

First Printing

National Library of Canada Cataloguing in Publication Data

Owen, David, 1939-
        Hidden secrets.

Includes bibliographical references and index.
ISBN 1-55297-565-7 (bound). ISBN 1-55297-564-9 (pbk.)

        1. Espionage. I. Title.

JF1525.I6O94 2002   327.12   C2001-901499-6

Publisher Cataloging-in-Publication Data (U.S.)

Owen, David.
    Hidden secrets / David Owen. –1st ed.
[224] p. : col. photos. ; cm.
Includes bibliographic references and index.
Summary: History of espionage around the world including descriptions of the technology used.
ISBN 1-55297-565-7
ISBN 1-55297-564-9 (pbk.)
1. Espionage -- History. 2. Espionage -- Equipment and supplies.
I. Title.
327.1209 21 CIP UB270.O94 2002

Published in Canada in 2002 by
Firefly Books Ltd
3680 Victoria Park Avenue
Toronto, Ontario
M2H 3K1

Published in the United States in 2002 by
Firefly Books (U.S.) Inc.
P.O. Box 1338, Ellicott Station
Buffalo, New York
14205

This book was designed and produced by
Quintet Publishing Limited
6 Blundell Street
London N7 7BH

Senior Editor: Laura Price
Editor: Ian Penberthy
Designer: James Lawrence
Picture Editor: Helen Stallion

Creative Director: Richard Dewing
Publisher: Oliver Salzmann

Manufactured in Singapore by Universal Graphic (Pte) Ltd.

Printed in China by Midas Printing Ltd.

Contents

Foreword

RIGHT Heavily-masked KGB defector Igor Gouzenko talking to members of the press after escaping to safety in the West.

BELOW Newspaper headlines show the release of British agent Greville Wynne as the result of an exchange with the KGB for their spy "Gordon Lonsdale."

As one of those in CIA who helped to defeat the enemy in the long conflict of spies between East and West known as the Cold War, I believe the lessons one can learn from this book illustrate some of the basic elements of the profession of espionage. In reading about espionage secrets, deception is and always was the key ingredient in these lessons.

The reader will find deception defined in the dictionary as a means to defraud, cheat or defame. But it is also defined as the means to gain strategic advantage and that is the service that spies provide for the well being of their individual causes. The professional spy must have a strong moral compass to know which are the good and bad lies otherwise he or she will get lost in what has been called the "Wilderness of Mirrors" not knowing which lie to believe. The spy must not lie for personal gain but must be skilled in mounting grand deceptions in order to protect the stratagems and secrets of his cause.

The strategies, tactics and tools of deception also have their roots in many of the older professions. Soldiers and generals have always used deception and illusion in the ways outlined by Sun Tzu in his book *The Art of War*. The reader will find these precepts discussed in this book as well.

Mystics and religious figures were early users of deception and illusionary techniques for their own brand of stagecraft in order to convince their followers of their ability to carry off miracles and therefore bind them even closer as true believers. A case could be made that the first intelligence operation in history was a covert action carried out by the devil disguised as a snake. As the first case officer he was able to convince Eve to betray Adam and partake of the apple.

Street performers and stage magicians have brought the principles of deception and illusion to a fine point over the centuries but primarily for our entertainment. Modern day entertainment media uses the latest in the state-of-the-art of technology to accomplish their illusions.

Prostitution is supposedly the world's oldest profession and its practitioners also employ some of the principles of deception and illusion.

Others believe that the business of spying is not the second oldest profession but it is the first. The point being made is that someone has always had to find out where the brothels are located and what is the current tariff.

Similarly the imaginations of intelligence officers and the budgets of intelligence agencies like the CIA, SIS and the KGB have driven the progress of technology over the years. Any gadget introduced in a James Bond movie probably already had a counterpart in reality. Society is the beneficiary of much of this real technology once it is transferred to the marketplace. Most are barely aware these high tech developments only exist because they were first invented to help deceive the enemy in days gone by. The reader will learn more about these things in this book. We can only guess what has not yet been revealed.

BELOW Keeping terrorists from committing atrocities like the crashing of an airliner into the Pentagon on September 11, 2001 calls for high-quality intelligence.

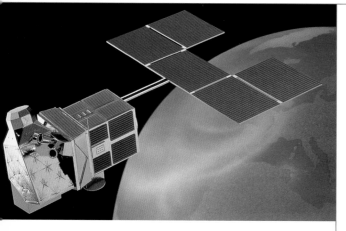

ABOVE Agents in space: *Helios* was the first defence intelligence satellite launched by the European Space Agency.

BELOW RIGHT The idea of the dedicated Nazi saboteur was a powerful threat in pre-war America as shown in this 1936 poster.

Today's technical means of intelligence collection have increased the number of questions and redoubled the need to have well placed human sources, called agents, working in place with access to the enemy's secrets. These human agents can best provide insights about the enemy's thoughts and intentions that intelligence collected by other means cannot.

If the agent gains access to a secret he must protect it with his life because he has probably already put his life on the line to acquire it. The task of any intelligence service is to keep their agents in place and alive as long as possible, to maintain timely and secure communications with them and get them out when their time is up. The means employed to maintain the security of these operations are known as tradecraft.

As a career technical operations officer in CIA, my colleagues and I were responsible for creating and deploying many of the forms of technical tradecraft to keep our agents secure. We made all the gadgets and were the real "Q" branch of James Bond lore. But unlike "Q," we also deployed the gadgetry in many cases and didn't leave it to the case officer to do by himself.

Much of the tradecraft still used in espionage operations is based on the tried and true techniques in use from the beginning of time. Even today espionage materials are passed from one spy to another by hiding them under a rock, or its equivalent, much like has been done since biblical times. These so-called "dead drops" are used because they work and are very secure if done correctly.

Disguise and illusion go hand and hand with deception and are more about how one must manage the operational stage, to know what is the enemies' point of view, and exactly who is the audience we are trying to fool. One must come to understand the audience's expectations and limitations. Then one must present them with an idea so they can follow it while you are simultaneously hiding the real purpose and action in the deception so it is never seen. Spies have employed these cloaking principles very successfully in very difficult operating environments and against overwhelming forces as we did in Moscow against the KGB. Many of these same

WALTER KAPPE, alias Walter Kappel

Walter Kappe is known to be connected with sabotage activities being promoted by the Nazi Government. He was born January 12, 1905 at Alfeld, Leina, Germany, and entered the United States on March 9, 1925. He filed application for United States citizenship at Kankakee, Illinois, in June, 1935. He is known to be a member of the German Literary Club, Cincinnati, Ohio, and the Teutonia Club, Chicago, Illinois. Kappe was an agent in the United States for the Ausland Organization and editor "Deutscher Weckruf und Beobachter", official organ of the German-American Bund. Kappe left the United States in 1937 and may return to the United States as an agent for Germany. This individual is described as fol-

principles are demonstrated in the case studies that follow in this book.

If every secret has a half-life then so does every spy. The typical life expectancy of a spy probably can be measured in months not in years although there were at least two cases during the Cold War where our agents were able to remain in place for ten and twenty years respectively. When it was time to bring them out of the cold we were able to warn them in time and cloak their escape long enough for them to vanish without a trace along with their families. These "exfiltration operations" are some of the most refined examples of good clandestine tradecraft. Some of these cases are discussed in this book. But so are the cases of some others that perished in the service of their cause.

Some spies were the most honorable of the great-unsung heroes and others were motivated to deceive for selfish reasons or a failed cause. Quite often right and wrong depends on one's point of view but as was said before, a strong moral compass helps one through the dark labyrinth that is the terrain of espionage. The lessons in this book should do likewise.

Antonio J. Mendez
Former Intelligence Officer *August 2001*

ABOVE LEFT Some spies are able to reach the highest levels of power in their target states: Günther Guillaume (with wife Christel) worked for the West German chancellor Willy Brandt, but spied for the East Germans.

BELOW Espionage technology of tomorrow—an artist's impression of the Aurora spyplane.

Introduction

RIGHT An example of the application of modern minaturized electronics to espionage technology: this miniature video camera is ideal for covert surveillance in homes or offices, and can be connected to standard television and video monitors for playing back the results.

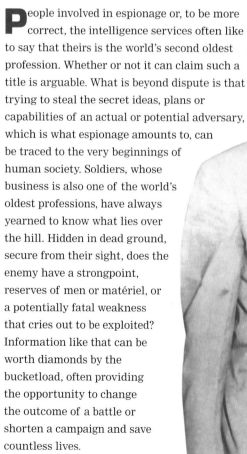

People involved in espionage or, to be more correct, the intelligence services often like to say that theirs is the world's second oldest profession. Whether or not it can claim such a title is arguable. What is beyond dispute is that trying to steal the secret ideas, plans or capabilities of an actual or potential adversary, which is what espionage amounts to, can be traced to the very beginnings of human society. Soldiers, whose business is also one of the world's oldest professions, have always yearned to know what lies over the hill. Hidden in dead ground, secure from their sight, does the enemy have a strongpoint, reserves of men or matériel, or a potentially fatal weakness that cries out to be exploited? Information like that can be worth diamonds by the bucketload, often providing the opportunity to change the outcome of a battle or shorten a campaign and save countless lives.

Even in times of international tension rather than real war, spies deal in potentially vital information. When countries are embroiled in negotiating treaties or arguing over claims to disputed territory, knowledge of a competitor's true negotiating position confers a priceless advantage. Any diplomat who is aware of his opposite number's ultimate fallback position is in the happy situation of a poker player who has glimpsed his opponent's hand. With facts like that at one's disposal, the rest is pure ritual and face-saving protocol. The outcome, in terms of power politics, is effectively already decided.

Yet, espionage and the information it generates have presented one problem, almost from the very beginning. In terms of human intelligence gatherers, who is the real spymaster? Is the agent who returns from enemy country really telling you what you need to know, or what your adversary wants you to believe? Even where inanimate forms of espionage, like signals, communications or image intelligence, are involved, a wide range of deceptions can be used to produce a false picture with the information.

This double-edged nature of the espionage weapon is one of the main reasons why spy stories retain their fascination in an increasingly sophisticated technological world. The idea of the double, triple or even multiple agent creates an image of shadowy half-truths and clever guessing games upon which hang terrible consequences. If an agent is suspected of working for the other side, the benefits from turning the tables on those who really employ him far outweigh the damage-limitation priorities of catching him and cutting off the channel of tainted information he represents.

By feeding the double agent with false information, you can reassure the enemy that he or she is doing an effective job, allowing them to relax in the belief that their true secrets remain hidden. In reality, the

LEFT Harry Gold, the North American spy arrested by Deputy Marshals following accusations of dealing with the British Atom spy, Dr Klaus Fuchs.

BELOW This artificial stone, made in Poland during the 1980s, has a hollowed centre in which messages can be left by agents without attracting the attention of casual passers-by.

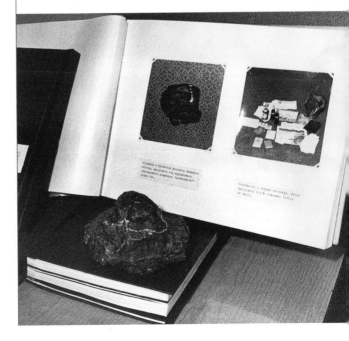

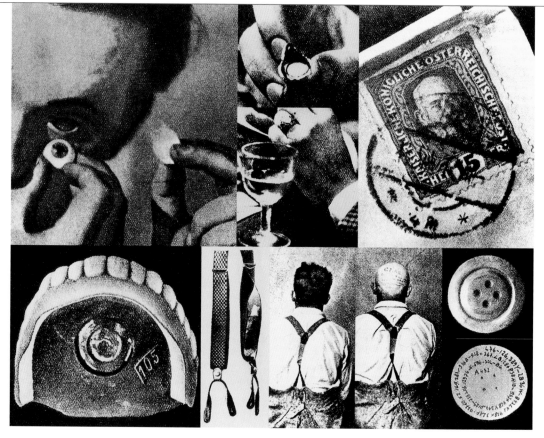

TOP LEFT An outwardly convincing glass eye, which can be used to hide microfilms or coded messages, and which would possibly escape attention even in a full body search.

TOP MIDDLE A gentleman's signet ring, with no external signs to reveal the hidden compartment large enough to conceal written messages.

TOP RIGHT Often agents could choose from a whole series of codes to pass on information—the perforations on this Imperial Austrian stamp have been cut to reveal the code used in the enclosed letter.

ABOVE LEFT A message written on a strip of thin cigarette paper and wound into a tight coil could be hidden in a cavity behind the teeth in this set of dentures.

ABOVE MIDDLE Some agents carried their messages in person—this Russian spy wrote information on his shaven scalp, which was hidden when the hair grew and only revealed when his head was shaved on his return.

ABOVE RIGHT A button containing a secret message, hidden by cutting the button in half, enclosing the message and then resealing it.

information the spy takes back to them represents what you want them to believe, while the planted information brought to you offers a priceless clue to what the enemy wants to hide. Unwittingly, the double agent has become a triple agent, and the true value of the information carried back and forth is switched between the two opposing sides.

Seen in this objective manner, the trade of the spy may appear to be a simple matter of searching for information and sending it or carrying it back to whoever controls the network of agents to which each spy usually belongs. In human terms, though, the cost of espionage is appallingly high. For reasons that vary from genuine ideological conviction or hatred of the target country to a greed for money or power, the spy is always under threat of exposure, capture and disgrace, with the prospect of a long prison sentence or even execution in the offing. To avoid these terrifying eventualities, he or she must trust no one and must learn to live a cover story to the full. Not for them are the normal compensations of career, family life and friendships, save under the most carefully controlled circumstances.

Even when the work is done, and the agent is pulled out of the front line to enjoy a well-deserved retirement in a country grateful for his or her efforts, the blessings can be mixed. Usually the host country is not one in which they were born or spent most of their lives, since the best agents are native to the societies where they trawl for information and where they can merge without detection. A country that was the focus of their loyalty and the source of their protection during the tense

ABOVE Guy Burgess, Kim Philby, Donald Maclean and Anthony Blunt, members of the infamous Cambridge spy ring that was active until the 1970s.

LEFT Aldrich and Rosario Ames stood trial for spying in 1985. He was found to have been a successful double agent.

operational years may be less inviting as a home without a job, in an alien society that often speaks a different language, even though their continued loyalty is not in question.

In the wider espionage picture, agents continue to play an important role, even though now the major discoveries tend to be made by more technologically based techniques. A human agent can reveal secrets that the most sophisticated intelligence gathering systems would not pick up. These may include the intentions of a nation's leaders in a particular campaign or crisis, the existence of dissident groups or individuals who could be recruited as additional agents to widen the sources of information, and shifts in

13

public attitudes that may be significant at times of heightened tension.

Nevertheless, today traditional spies have to earn their keep in an immeasurably more complex and capable world than their predecessors. Since the invention of radio and radar, reconnaissance and intelligence gathering aircraft and satellites, and systems for intercepting and decrypting communications, the secrets revealed by the most successful of human agents contribute to a much wider and more detailed picture assembled from all these varied sources.

Different chapters of this book are devoted to SIGINT (signal intelligence, obtained from the interception and decrypting of messages, and the analysis of traffic), ELINT (electronic intelligence, involving remote sensors and traffic surveillance), FALSE INTELLIGENCE (employing deception, subversion, misinformation and the use of double agents) and IMINT (image intelligence, including video

intelligence and airborne information gathering using technology like infrared line-scan—IRLS—and satellite surveillance).

This book encompasses all the aspects of intelligence and information gathering, and the efforts that countries and their counterintelligence organizations employ to keep their secrets safe from prying eyes. Yet, there remains one positive aspect of espionage, which can act across national boundaries and help to make the world a little safer. So many wars have broken out because one nation misunderstood another's capability, attitude or intention. At one time, revealing the hidden secrets could convince a potential aggressor that an attack was feasible. Later, in the context of the Cold War, the balance of terror required that both East and West remained in no doubt at all of the opposition's ability to ensure MAD— Mutual Assured Destruction— and thereby deter the pressing of the nuclear button. In this respect, the value of the spy as

TOP The two faces of Juan Pujol, codenamed Garbo, one of the most successful double agents ever.

LEFT The directors of the Russian Communist Cheka secret police organization in 1918.

TESTIMONIANZA FORNITA

...tità: Carta d'identità n° 131 rilasciata dal Comune di Termeno
1'11-VI-1948
(Documenti personali presentati)

...grazione: Permesso di libero sbarco n° exp.231489/48
(indicare se avverrà tramite un Comitato responsabile. Designazione dell'Autorità. Nom. di registrazione)
...rtenza col Piroscafo ANNA "C" nella prima metà di Giugno

...ivatamente (indicare promesse di visto ottenute):

CONNOTATI

Capelli: castani
Occhi: celesti
Naso: regolare
Segni particolari:

...ronta digitale
...ollice destro)

...isto per l'autenticità delle dichiarazioni, fotografia, firma e im-
...ronta digitale del Sig.   Klement Riccardo
...irma e timbro dell'Autorità: P. Drmohr Edoardo
...uogo e data: Genova 1/6/1950
(pregasi apporre il timbro anche nella fotografia)

...ta 10.100 bis N.   100940   Validità   un anno
...cessa   a   Genova   il   1/6/1950
"   "   "   "
...segnata a   il

Firma del richiedente ⟩——▶

ABOVE The forged Argentinan identity papers of Nazi official Adolf Eichmann that led him to be able to evade capture and trial.

a channel of communication to prevent fear of a preemptive strike was supreme, and reluctantly acknowledged by both sides.

Today, with a greater number of less powerful potential adversaries, perhaps the place of the agent and the secrets that remain the target of his or her operations have reverted to the espionage traditions of the past. The world of 21st-century espionage, examined in the closing chapter, is challenging and so far largely unpredictable. Yet, one factor is certain—revealing the hidden secrets of tomorrow will continue to engage the best minds and the bravest agents, who will be faced with the most determined efforts of those who guard them. For the second oldest profession, the future promises to be every bit as busy as its long, varied and colorful past.

ABOVE Hollow pen shaft used to hide and transport papers and codes.

RIGHT Francis Gary Powers under trial by the USSR Supreme Court's Military Collegium.

The Early Days

ABOVE A graphic 16th century Persian depiction of terror and bloodshed as Genghis Khan's army storms a fortress, with its defenders facing massacre, the secrets of their defences betrayed by Mongol spies.

With such a secretive and long-established trade as espionage, discovering its origins is all but impossible. In any case, the earliest spies were hardly spies at all, in the modern sense of the word. Reports would be sent back by a nation's diplomats, who would be in close touch with all shades of opinion and all levels of government in the country to which they were accredited. Visitors to foreign lands for purposes of trade or simply curiosity were often questioned upon their return, adding to the overall picture that showed whether a nation was a potential ally or adversary, or whether it would help to maintain the existing balance of power.

With these largely amateur sources of information, there were limits to the discoveries that could be made, and the secrets that other countries wanted to keep hidden usually remained so. The realization that employing agents who could merge into the society of the target country to ferret out these secrets was a much more effective way of seizing information occurred at widely differing times in different parts of the world. In China, it was the military sage Sun Tzu who, more than two-and-a-half millennia ago, understood the priceless value of knowing what his opponents would be likely to do in a given situation, and also the power gained by being able to pass on misleading information in return.

Sun Tzu believed that the ideal way to defeat an opponent was to avoid battle altogether, and to use deception and clever maneuvers to confuse, distract and ultimately triumph over the enemy:

"The enemy must not know where I intend to give battle. For if he does not know where I intend to give battle, he must prepare in a

great many places. And when he prepares in a great many places, those I have to fight in any one place will be few. And when he prepares everywhere, he will be weak everywhere."

Sun Tzu made widespread use of spies in his quest for knowledge about real and potential enemies. He believed that secret operations were "essential in war…an army without secret agents is exactly like a man without eyes or ears…of all those in the army close to the commander, none is more intimate than the secret agent."

But spies had power of their own—by passing on false information they could work for the enemy, which meant it was essential to recruit only the very best people: "The first essential is to estimate the character of the spy to determine if he is sincere, truthful and really intelligent…Afterwards, he can be employed…Among agents there are some whose only interest is in acquiring wealth without obtaining the true situation of the enemy, and only meet my requirements with empty words. In such a case I must be deep and subtle."

What kind of information should the clever secret agent look for? Sun Tzu quoted several telltale signs that told their own story:

"Dust spurting up in high straight columns indicates the approach of chariots. When it hangs low and widespread, infantry is approaching…when half an enemy's force attacks and half withdraws, he is attempting to decoy you…when his troops lean on their weapons, they are famished…when drawers of water drink before carrying it to camp, his troops are suffering from thirst…when the enemy sees an advantage but does not seize it, he is fatigued…when birds gather above the camp sites, they are empty…when at night the enemy camp is clamorous, he is fearful…when his flags and banners are constantly on the move, he is in disarray…"

Secret messages in Greece and Rome

But if Sun Tzu laid the foundations of espionage, other aspects of the craft were already being developed by the early civilizations of Europe. During the fierce and often bloody rivalries between the city-states of classical Greece, Sparta played a leading role and had a formidable military reputation. Like the Chinese warlords, the Spartans realized that successful military operations depended on the ability to send information between their agents and their generals without the enemy being able to read those messages and benefiting from their contents. To this end, they invented a device called the "skytale," which was a standard baton wrapped with a strip of papyrus, leather or parchment. The message was written on the wrapping, then the strip was removed and passed to the messenger. Unwound, the strip only showed an apparently random set of letters, which made no sense at all. Only when it was rewound around a baton of the same diameter as the original skytale could the message be deciphered.

The Romans also made use of spies and secret ciphers. Julius Caesar encoded his letters to Cicero and other friends simply by shifting the letters three places along in the alphabet, and to this day this type of cipher is known as a Caesar alphabet. Later, the Arabs

carried out the first logical study of how to break ciphers by using the frequency with which standard letters appear in any piece of writing in a particular language; this can provide a route to a hidden message. In English, for example, "e" is the letter that appears most frequently in general language, followed by "t" and so on.

Even the most cruel and despotic conquerors found spies useful in many different ways. The fearsome Genghis Khan, who led the Mongols on their sweeping conquests out of Central Asia and into Europe during the 13th century, employed agents for two purposes. Apart from ferreting out information that told him of an enemy's strengths and whereabouts, their plans and their weaknesses, he used them to generate fear and defeatism among those he was about to attack. When a city defended itself, his agents spread rumors that a quick surrender would bring merciful treatment, although the truth was vastly different. When a city opened its gates to the Mongols, they massacred most of the inhabitants; when it defended itself with spirit, they killed them all.

Spies of the Reformation

Espionage came into its own through the power politics of Renaissance Europe. As the trading republic of Venice established its authority over the Eastern Mediterranean during the late 15th century, control was maintained by a shadowy and sinister group called the Council of Ten, which ran a large and efficient secret police force, with agents and contacts reporting from within the republic's major competitors. The Council's cipher secretary, Giovanni Soro, was appointed in 1506 and charged with the task of deciphering secret messages captured from the spies and messengers of Venice's rivals. His achievements led to him being commissioned by the Papacy to unravel messages seized by

their own secret agents. So successful were Soro's efforts, as were those of his overseas competitors, like the French lawyer François Viète, that the Council regularly changed its own secret ciphers to prevent Venetian messages from being read in their turn.

In Elizabethan England, spies were used to expose plots against the life of the monarch herself. When a former page of Mary Queen of Scots, Anthony Babington, planned the assassination of Elizabeth in 1586 and the restoration of Roman Catholicism, the conspirators smuggled coded letters into the house where Mary was held a prisoner by hiding them in beer barrels. But the man whom Babington recruited as a messenger and who devised the smuggling routine, Gilbert Gifford, was a double agent under the orders of Sir Francis Walsingham, who ran Elizabeth's espionage service. Walsingham relied on a wide-ranging group of spies from all levels of society, including the poet and playwright Christopher Marlowe, who was a rival of William Shakespeare.

Walsingham's expert, Thomas Phelippes, deciphered copies of the letters, and when he revealed a message that showed Mary supported the murder of Elizabeth and an invasion by her ally, King Philip of Spain, he drew a gallows symbol on the parchment. It was Mary's death warrant, but first Walsingham wanted to catch the conspirators. He ordered Phelippes to add a forged paragraph to one of Mary's letters to Babington, asking for the names of the men involved.

ABOVE Spies of Queen Elizabeth I **(CENTRE)** revealed that plotters supporting Mary Queen of Scots **(ABOVE LEFT)** planned her assassination, in alliance with King Philip II of Spain **(ABOVE RIGHT)**.

OPPOSITE PAGE The Roman statesman Marcus Tullius Cicero, one of the group of friends of Julius Caesar who communicated through coded letters using the "Caesar Alphabet", a transposed letter cipher.

BELOW Cardinal Richelieu founder of the Cabinet Noir intelligence organization to protect the Kings of France from plotters and conspirators.

Before the reply was received, Babington decided to visit his collaborators on the continent. Walsingham engineered a mix-up with his passport, so that when Babington sought the former's help to solve the problem, he could be arrested. Unfortunately for Walsingham, Babington realized what was happening while dining in a tavern with one of Walsingham's agents, and managed to escape. However, within a matter of weeks, Babington and the others involved in the conspiracy had all been caught, and they and their mistress were sent to the block for treason.

England's greatest rival, France, also developed a formidable and successful secret service from the body set up in the reign of Charles V during the 14th century. But it was

Cardinal Richelieu who was able to develop the first effective French espionage organization, drawing on his position in the Catholic Church which, at that time, maintained the most efficient espionage setup in the world. In 1620, he established a network of advisors on intelligence matters, which was called the Cabinet Noir. Their purpose was to evaluate intercepted letters and reports from spies in every part of the kingdom, who were watching for discontent and plots that could threaten the monarch. This secret service was managed by another Catholic priest, François le Clerc du Tremblay, usually known by his cover name of Father Joseph.

The achievements of Richelieu and his spies had a major effect on European history. He kidnapped an English agent called Montague, who revealed the efforts of the Duke of Buckingham to meddle in French affairs.

TOP LEFT Sir Francis Walsingham head of Elizabeth's secret service, whose agents revealed the existence of the Babington Plot and the involvement of Mary Queen of Scots, which led to her execution and the hanging, drawing and quartering of the plotters.

TOP RIGHT The Protestant King of Sweden, Gustavus Adolphus was persuaded to enter the Thirty Years' War by Richelieu in a complex plot to gain France the province of Alsace.

LEFT The Babington Plotters meet their fate and are hanged before being drawn and quartered for treason.

*In quo quis peccat
In eo punitur.*

Early spy technology: secret inks and secret signals

Compared with the techniques of modern espionage, the agents of the 18th century had a much more difficult task to pass on their information through a country ravaged by revolution. Consequently, they tended to place great reliance on invisible ink, using it to write between the lines of otherwise innocent papers. The ink was made readable by applying the right treatment.

The Culper Ring spy network, reporting to George Washington's intelligence chief, Benjamin Tallmadge, from the British headquarters in New York in 1780, used a preparation called Jay's Sympathetic Stain. When written on white paper, the message disappeared as the ink dried and could only be revealed by brushing the paper with the right reagent.

The network was based in Tallmadge's home town of Setauket, on the shores of Long Island. Messages to the New York agents were written on a sheet of paper included at a prearranged position between blank sheets of a regular order for writing paper, collected from a New York store by one of the agents, Austin Roe. When Roe returned to his cattle on land belonging to a farmer at Conscience Bay, he would place the sheet with the hidden message in a box buried in a corner of the pasture. Another agent would retrieve it and take it to a rendezvous with Caleb Brewster's whaleboat.

Brewster's arrival would be signaled by a black petticoat fluttering on agent Anna Strong's clothesline, while the number of handkerchiefs hanging alongside it would identify the cove where the boat lay. Once the message had been delivered, Brewster would row back to the Connecticut shore and pass it to Tallmadge, who sent it by relays of horsemen to Washington's headquarters on the Hudson river.

The system worked well until late in June 1780, when British cavalry all but captured Tallmadge at the hamlet of Poundridge, and many of his papers were lost. They revealed the use of secret ink (although not its composition, nor how to reveal the hidden message) and the names of several agents. To guard against future messages being discovered, Tallmadge reinforced them with codes. Places and people were given three-figure numbers—Austin Roe was 724, Washington 711, New York 727 and Long Island 728, for example.

In any event, the days of the Culper Ring were numbered. Several agents resigned, and messages ceased to be passed until later in the summer of 1780, when the spies within New York came upon vital information that was essential to relay to Washington. This was written in the Sympathetic Stain ink on the back of an ordinary business letter to a Colonel Benjamin Floyd in Brookhaven. Austin Roe carried the letter, which explained that the goods ordered by the colonel were not in stock, which accounted for Roe riding empty-handed.

Washington's headquarters were expecting French reinforcements at Newport, Rhode Island, but the spies revealed that Clinton's troops were about to leave New York for Newport, supported by ships of the Royal Navy under Admiral Graves. Washington hurriedly drew up a letter containing plans for a full-scale attack on New York with a much stronger army than he actually possessed. This was placed with other correspondence and given to a double agent, who passed it to the British, explaining that he had found it on the road. Once they read the false letter, the British ordered Clinton's forces to return, while the fleet sailed back to New York, leaving the French to land their troops unopposed.

ABOVE Benjamin Tallmadge, George Washington's chief of intelligence and the mastermind behind the Culper spy ring.

Richelieu went on to persuade the King of Sweden, Gustavus Adolphus, to enter the Thirty Years War on the Protestant side, which so distracted the Catholic alliance that it enabled France to seize the rich province of Alsace, and later weakened Spanish power by inciting rebellions in Portugal and the province of Catalonia.

Spies across the Atlantic

For all their usefulness in periods of rivalry and tension between governments during peacetime, spies and agents are most vital in times of war, when their courage or deceit can influence the outcome of the bloodiest or most hard-fought campaign. At no time was this more obvious than during the war of the American Revolution, as a result of which the American colonies won their independence from England. One of George Washington's most valuable spies was the double agent Sergeant John Honeyman, who had fought with the American leader in the British Army during the Indian War in 1763, and who passed on information that enabled the Americans to

cross the Delaware and capture Trenton in the fierce winter of 1776. At a time when the American cause had been all but extinguished by a succession of defeats and pursuits, combined with hunger, bad weather and a drop in morale, this sharp change in fortune revitalized their campaign.

Later, during the conflict around New York in 1780, George Washington's intelligence chief, Colonel Benjamin Tallmadge, maintained a network of spies, called the Culper Ring, within the city, when this was the headquarters and primary base of the British forces. Recruited from traders and craftsmen who worked for the British Army, it relayed vital information across Long Island Sound to Washington's headquarters (see sidebar). Nevertheless, loyalties crossed the fighting lines in both directions, and one of the highest placed British agents proved to be no less than General Benedict Arnold, while Edward Bancroft, secretary to Benjamin Franklin in Paris, was reporting directly to London.

For all that, Washington was able to pull off another dramatic coup against his British opponents. Warned by his spies that the enemy was about to move with naval support against rebel forces around Newport, Rhode Island, the American commander was able to send a double agent with a false message reporting an imminent American attack on New York itself. Although the truth was very different— Washington's forces were far too weak to mount a frontal assault on the city—the information from such an apparently reliable source was enough for the British commander to retreat within the city's defenses, leaving the way clear for Washington's French allies to land at Newport without any opposition at all.

TOP LEFT Father Joseph was the cover name for Richelieu's associate who ran the Cabinet Noir—former Catholic priest Francois le Clerc du Tremblay.

OPPOSITE George Washington was able to make good use of undercover agents and secret intelligence in the campaigns of the Revolutionary War.

Washington's double agent, Sergeant John Honeyman

By the bitterly cold December of 1776, George Washington's army had been hammered by the seasoned British regulars and their Hessian mercenaries in a series of reverses at Long Island, White Plains and Fort Washington. Washington's men had to retreat across New Jersey to the Delaware River crossing, where they won a respite by seizing or destroying all the available boats. Lacking tents and blankets, and staying alive on scraps of bread, they watched their Hessian opponents observing them from the village of Trenton.

The British were waiting in secure winter quarters until warmer weather would allow them finally to crush the rebellion. On December 18, Washington wrote to his brother, John Augustine Washington, telling of his despair for the future. Yet in a matter of days, the intervention of a 46-year-old Ulsterman and former sergeant in the British Army would change his fortunes completely and boost the hopes of the rebel cause.

John Honeyman was a weaver, but more than 10 years before he had served alongside Washington in the British Army, fighting the French and the Indians. On December 22, 1776, he was searching for stray cattle along the river when he was spotted by two American cavalrymen, who seized him as a suspected spy. He was dragged before George Washington, who dismissed his aides while he interrogated the prisoner.

In fact, Washington and Honeyman had met a year-and-a-half earlier, with rebellion in the air, when Honeyman had volunteered to serve as a spy against the British. He was given a new cover story as a cattle dealer and butcher, which gave him a reason to trade with the army. To preserve his safety, only Washington knew his true identity, and they could only communicate by Honeyman allowing himself to be captured by American soldiers.

This time, he had vital information. Both British and German units across the Delaware were off guard, distracted by the approaching Christmas holiday. A well-timed attack might succeed, particularly if the British were convinced of their opponents' pitiful condition. That night, Washington ordered the spy locked in the guardhouse, to await court-martial the following morning.

BELOW Washington crosses the Delaware at the head of his retreating army at a moment of great peril for the American cause.

Except that John Honeyman vanished during the night, during the chaos caused by a fire in a nearby hayrick. He crossed the river back to British territory and reported to the commander, Colonel Rall, that he had been seized by the Americans, accused of spying and locked in the guardhouse overnight. He described his escape, emphasizing that the enemy was on the verge of mutiny and incapable of any offensive action at all. Rall was delighted to have his own estimates confirmed and returned to the festivities.

While feigning anger at the spy's escape, Washington prepared his army for a recrossing of the Delaware and an attack on Trenton on the evening of Christmas Day. Even then, Rall was warned by his own agents that the Americans were preparing to fight, but he refused to leave a card game to deal with a threat that contradicted his own information. When Washington's men attacked in a blizzard, they achieved total surprise: Trenton was taken, Rall was killed and the garrison surrendered. The rebellion was revived with a genuine victory at the cost of two officers and two enlisted men wounded. John Honeyman faded into obscurity, dying in his bed at the age of 93.

BELOW Washington watches his men crossing the Delaware river to win a vital respite from the British pursuit.

Colonel Alfred Redl, master spy of Imperial Austria

Colonel Alfred Redl was a very successful officer in the army of the Austro-Hungarian Empire when, in 1901, he was appointed as deputy chief of the Evidenzburo military counterespionage organization. As such, his main target was Russia, and for four years he worked hard to convince the authorities that he was an able and dedicated intelligence expert. Unfortunately, his double life among Austria's gay community enabled the Russians to blackmail him, and he became a double agent within a year of his appointment.

One of the measures Redl had introduced was the censorship of mail sent through the Imperial postal system. In 1905, he was promoted intelligence officer of the VIII Corps, based in Prague, and replaced at the Evidenzburo by Maximilian Ronge. Redl continued to work as a double agent for more than a decade and, thanks to lavish payments by his Russian spymasters, lived well beyond his official means.

Nevertheless, doubts began to grow, and Redl must have been aware of them. Two large cash payments, totaling 14,000 Austrian crowns, had been sent by his Russian controllers, addressed to a fictitious name, to be picked up from the central Vienna post office. Redl was careful not to collect them until he was sure it was safe. Unfortunately, the consignments had been reported to Ronge, who noticed that the postmark indicated they had been posted in a small town on the border between Germany and Russia. He ordered a watch to be kept at the post office, but no one collected the money for almost three months.

When the packages were picked up on May 25, 1913, the police missed the man by seconds, and he and his packages disappeared in a cab. Eventually, they managed to identify the cab and discovered that the passenger had been taken to the Kaiserhof café, where a second cab had taken him on to the Hotel Klomser. But he had left a gray suede cover from a pocketknife in the back of the cab. When the police checked at the hotel, they found that the cover belonged to Colonel Redl, who was registered there as a guest. Checks at the post office revealed that the false signature given for the packages was recognizable as Redl's handwriting.

By this time, Redl realized that he was being followed, and he ran from the hotel, followed by police detectives. Twice he pulled papers from his pocket, tore them up and threw the fragments away, before returning to the hotel. The detectives collected the scraps of paper, which included a receipt for sending money to a German cavalry officer, and receipts for registered letters to addresses in Brussels, Warsaw and Lausanne. These addresses appeared in Redl's own counterintelligence files as those of foreign intelligence services.

The situation was reported to the Austrian secret service and the army, and four officers were sent to confront Redl. He told them that all the information they needed was to be found in his house, and asked for a loaded revolver. The officers went to wait in a nearby café with a view of the hotel entrance. At five o'clock, they checked his room and found his body; he had shot himself through the head.

In his time, Redl had sold the mobilization plan against Russia, details of a network of fortresses along the Galician border with Russia, codes and, almost certainly, the identities of Russian agents working for the Austrians, thereby cutting off Austrian intelligence on their most formidable adversary.

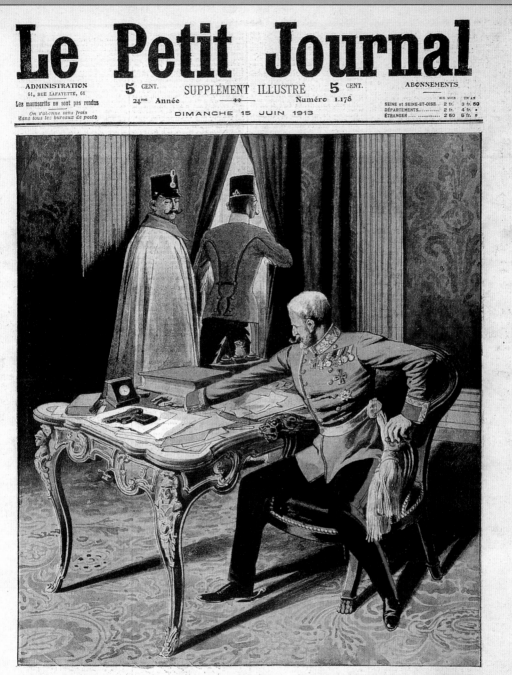

Le Petit Journal

ADMINISTRATION
61, RUE LAFAYETTE, 61

Les manuscrits ne sont pas rendus

On s'abonne sans frais
dans tous les bureaux de poste

5 CENT. 24me Année

SUPPLÉMENT ILLUSTRÉ

5 CENT. Numéro 1.178

ABONNEMENTS

SEINE et SEINE-ET-OISE.. 2 fr. 3 fr. 50
DÉPARTEMENTS............ 2 fr. 4 fr. »
ÉTRANGER .... 2 50 5 fr. »

DIMANCHE 15 JUIN 1913

LA TRAHISON DU COLONEL REDL EN AUTRICHE
Comment le coupable s'est suicidé

OPPOSITE PAGE Colonel Alfred Redl in the uniform of the Imperial Austrian Army.

ABOVE How a French magazine portrayed the suicide of Redl, when his treachery had been discovered.

The woman who never was

ABOVE Madame de Pompadour, mistress of the French King Louis XV **(OPPOSITE TOP)** plotted against the Chevalier d'Eon (portrayed here dressed half as a man and half as a woman) **(BELOW RIGHT)** resulting in his being recalled to France from the Court of the Russian Tsarina Elizabeth **(OPPOSITE BOTTOM)**.

One of the most unusual and remarkable of French agents, the Chevalier d'Eon de Beaumont, was recruited by King Louis XV's Secret Correspondence espionage organization in 1754. Then aged 26, de Beaumont had unusually fine features. Moreover, his mother had dressed him in women's clothing until his early teenage years. Although a skilled swordsman and otherwise entirely masculine, nevertheless he was able to pass himself off as a woman with ease, and he was sent to the Court of the Tsarina Elizabeth at St. Petersburg, posing as Mademoiselle Lia de Beaumont.

His orders were to use his influence to persuade the Russians to switch from their pro-British alignment to one more sympathetic to France. He took with him a letter from the King to the Tsarina, and his sympathetic advice succeeded in persuading her not to sign a treaty with the British, but to establish close relations with Austria and France. When he revealed his true identity, the Tsarina offered him an important court position, but he returned to France to be honored as a Chevalier of the Grand Cross.

He was sent, in male attire, to Russia when Catherine inherited the throne and resumed a pro-British stance, and once again he was able to change official policy. At the age of 34, he was sent to England to work as secretary to the French ambassador, but his secret purpose was to seek out the best routes for an invasion of England and seizure of the capital. While he was away, the King's mistress, Madame de Pompadour, plotted against him, and he was recalled to Paris to face attempts on his life.

He fled to London, where he was to live for the rest of his long life, dying in 1810 at the age of 82. Several times the French tried to persuade him to undertake more missions, but he always refused. On one occasion, they hid a sweep in the chimney of his house in London, with orders to groan and persuade de Beaumont that the house was haunted. Expecting him to report this to the authorities, they were confident that he would be sent to an asylum. Instead, de Beaumont drew his sword and called up the chimney, threatening to kill whoever was hidden there unless they gave themselves up. The sweep confessed that he had been hired by the French, and de Beaumont retaliated by publishing scandalous stories of the private life of Louis XV.

The French tried again, sending another agent with the codename "Norac" to persuade him. This was the playwright Caron de Beaumarchais, author of *The Barber of Seville* and *The Marriage of Figaro*, but this attempt also failed. De Beaumont returned to France once more, when Louis XVI had succeeded to the throne and the Secret Correspondence had been disbanded, but later he resumed residence in England.

Treachery at the top: General Benedict Arnold

The American General Benedict Arnold was a man with a grievance. He had fought in the attack on Fort Ticonderoga in May 1775, with Ethan Allen, and had tried to attack the strongly fortified city of Quebec with a force of 700 men, but the attack had failed and he had been wounded severely. Promoted to the rank of brigadier general, he commanded a small squadron of ships on Lake Champlain and defeated the British in October 1776, but was passed over for promotion to major general in February 1777, and Washington had to persuade him not to resign his commission. In May, he was belatedly promoted after repelling a British attack on Danbury in Connecticut, but remained junior to officers of lesser experience who had been promoted before him. Only after another victory against the British and commanding the advance guard at the Battle of Saratoga was he given full seniority.

By then, Arnold had been badly wounded and had been appointed commander of Philadelphia, where he mixed with loyalists who opposed the war with Britain. In 1779, he lobbied for the command of West Point, which Washington was prepared to give him, but he was asked first to command the left wing of the army, which was about to disrupt the British campaign in Rhode Island, intended to isolate the rebel army from its French allies.

Arnold declined, complaining that his wounds made this impossible, but by then he had a more pressing reason. In May 1779, he had begun a secret correspondence with Major John Andre, adjutant to the British commander, Sir Henry Clinton. Andre and a Colonel Robinson sailed up the Hudson River in a British sloop of war, *HMS Vulture*, to deliver letters to Arnold at West Point. Arnold insisted on the British officer changing from uniform into civilian clothes, but on his way back to the ship he was arrested as a spy.

When searched, Andre was found to be carrying plans of the West Point fortifications, details of the weapons and stores there, and a report of General Washington's latest council of war. He was charged with espionage, found guilty, and executed two days later. Meanwhile, Arnold escaped on a British ship, to find that leaving Andre to his fate had made him as unpopular with the loyalists as he was now with the rebels. He retired to England, where he died in 1801 at the age of 60, ostracized and unemployed.

OPPOSITE BOTTOM RIGHT
Benedict Arnold had taken part in the attack on Fort Ticonderoga with Ethan Allen **(OPPOSITE FAR RIGHT)** and was badly wounded at the Battle of Saratoga **(OPPOSITE TOP)** before beginning a secret exchange of information with officers on the staff of the British commander Sir Henry Clinton **(ABOVE)** and finally revealing his treason to his shocked wife **(OPPOSITE)**.

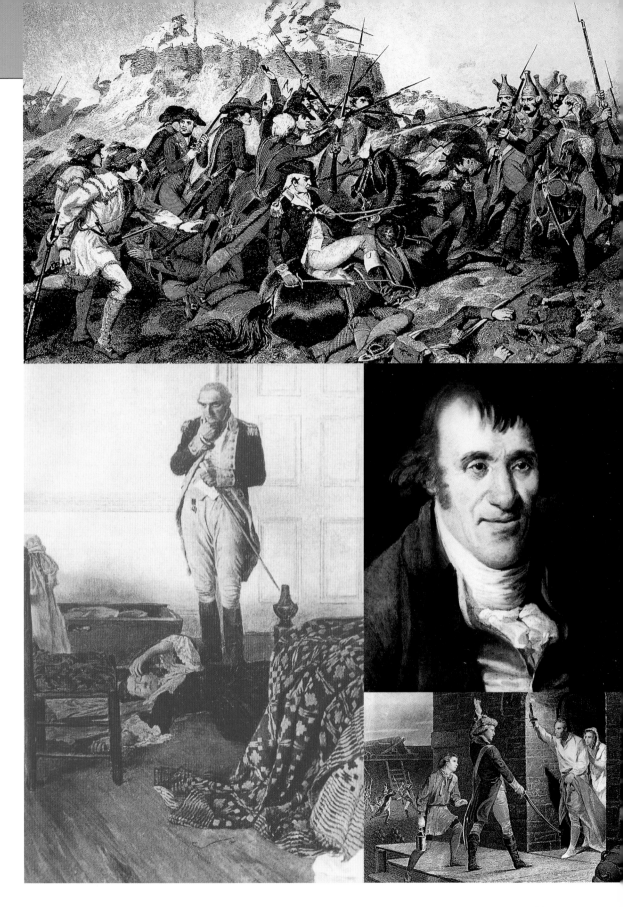

Spies in the Great Wars

ABOVE Major Henry Hardinge, chief of intelligence to the Duke of Wellington **(OPPOSITE TOP)** commander in chief of the British Army fighting the forces of the French Emperor Napoleon **(OPPOSITE BOTTOM)** in Spain.

Compared with plots to overthrow a monarch, seize a country or win a victory against a determined and successful enemy, many of the spies of the 19th and 20th centuries lacked an obvious vital target to make the loneliness and danger of their work appear worthwhile. The difference, of course, was war. In most cases, spies were seen as a necessity during peacetime, even if the true reasons for employing them owed more to suspicion of a country's allies and paranoia about its potential opponents.

Once hostilities broke out in a major conflict, spies became as essential as any other potential battle-winning weapon. Ultimately, knowing what a possible opponent planned in peacetime was far less important than finding out what that same opponent intended to do in war.

Even in the Napoleonic Wars, spies and intelligence experts were able to influence great events. The Duke of Wellington, commanding Britain's army in Spain against the forces of Napoleon, relied on information supplied to him by his Chief of Intelligence, Major Henry Hardinge. In the days when a general's knowledge of the enemy was often limited to what he could actually see, advance information of his opponent's whereabouts, and in which direction his forces were marching, became essential.

Napoleon had won an empire by his skill of maneuver. He was able to spread his forces over miles of countryside to disguise his underlying intentions. Only when he had chosen the point at which to strike did he assemble his forces quickly enough to win a decisive victory against an opponent who may have been stronger overall, but whose forces were fatally dispersed.

Spies in Spain

In Spain, Napoleon faced what was to become a fatal disadvantage. Because the French had turned from being allies of the Spanish to become their oppressors, most Spaniards were willing to help break the French yoke. Guerrilla units hiding in the mountains ambushed French columns and supply trains, while farmers and traders reported where French units were billeted and tracked their movements across the countryside. Those Spanish people who were employed by the French Army as cooks, cleaners, servants and laborers often saw and heard much more. Messages were passed to

friends and relatives in the countryside, and eventually this information reached Hardinge, who used it to develop a coherent picture of what the French were planning and when those plans might be carried out.

Time and again, this enabled Wellington to outwit the generals whom Napoleon sent against him, who disposed their forces according to their leader's orders, but who lacked his own particular genius. Nevertheless, the French were able to pour more and more men and resources into the Spanish conflict, and at times Wellington's more limited forces were close to defeat, being forced to retreat behind strong defenses in Portugal. When one of Wellington's officers, Major Colquhoun Grant was taken prisoner by the French in 1811, and held at Marshal André Masséna's headquarters, this seemed merely to be an additional disaster. In

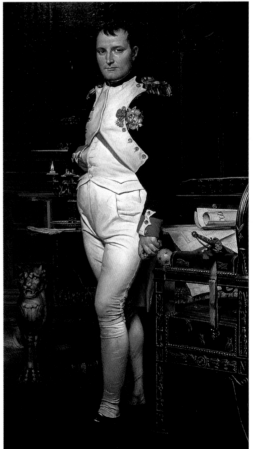

ABOVE French officers of Marshal Andre Massena revealed vital information to their prisoner, Wellington's agent Major Colquhoun Grant.

fact, while Grant was being entertained by his captors, in accordance with the etiquette of the time, he overheard conversations between French officers that referred to Napoleon's planned invasion of Russia and the diversion of troops from Spain to take part in the attack.

Tilting the balance

Grant managed to make contact with one of Wellington's spies within the French encampment and passed on the information. With this knowledge at his command, Wellington was able to play a waiting game. Since the French troops were not reinforced, and losses not replaced, the balance of power in Spain began to shift. When 30,000 troops of the French Army in Spain were ordered to join the forces earmarked for Russia, Wellington's opportunity for victory was confirmed.

Hardinge paid his Spanish agents well for any documents they were able to steal from the French, but one of them returned from the French headquarters with a prize that was beyond price, in the shape of Marshal Masséna's codebook. This enabled Hardinge to decipher captured dispatches that revealed the detailed French battle plans when Wellington marched his troops back into Spain. The Duke's final Spanish victory, at Vitoria on June 21, 1813, paved the way for the invasion of France itself.

Trained observers who realize the value of the information they uncover are worth their weight in diamonds to those who control them. Some agents may pass on any information, no matter how unreliable or unconfirmed, to justify their employment. On the other hand, there is always the danger that an agent may not, from his or her own limited perspective, appreciate the true value of a snippet of information. The truth is that sometimes the most apparently trivial detail may be the final fact needed to verify a particular development.

This accounts for another trend in modern espionage, arising from the employment of double agents and misinformation. Since all spies operate in conditions of secrecy in other countries, it is difficult for those controlling them to be certain of their loyalty. Even if their allegiance is not in dispute, the other side may have identified their role and be feeding them with false information. So all espionage involves a complex balancing act. Is the information being sent back by agents true? If it is, it may be very valuable. Is it false? In this case, it is less valuable, but still worth considering as an indication of what the opposition might want to conceal. In the latter situation, is it possible to lull the opposition into a sense of false security by releasing misleading information in return, to suggest that the bait has been swallowed?

The importance of detail

Even the smallest of details are important in the overall intelligence picture, particularly in wartime, when travel within the target country will be difficult or impossible. The late Professor R.V. Jones, who played a crucial role in wartime British intelligence, once explained how he decided that references to German weapons research at Peenemünde, on the

Baltic coast, were genuine. The discovery from a completely different source of German wartime priorities in the allocation of gas and oil showed that the Peenemünde establishment ranked very high. Reasoning that no one laying a false information trail would cover their tracks to that extent, Jones concluded that Peenemünde was worth further investigation.

What would have happened if the evidence had been lacking and the intelligence chiefs had concluded that the story was a fabrication? Not only would the selfless courage of the agents have been wasted, but also the Germans' development of potentially war-winning weapons would have continued unhindered. Allied efforts to discover what was happening in that remote research establishment on the coast of northeast Germany would not have been made, nor would they have attempted to stop the program by bombing the laboratories and the houses of the scientists who worked in them. At the very least, the onslaught of the V1 cruise missiles and V2 ballistic rockets would have burst upon the Allies as a totally unexpected, and therefore much more damaging, campaign.

In World War I, German U-boats were sinking large numbers of Allied merchant ships, which were carrying much-needed food and raw materials to Britain. These ships were sailing without the protection of convoys and escorts, and it was all too easy for a submarine

BELOW A terrible new weapon makes its appearance, and threatens to cut off the Atlantic supply route: a German U-boat sinks the liner *Fabala* in 1915.

to sink one without using a scarce and expensive torpedo. Instead, the submarine invariably surfaced and sank the ship with its deck gun.

In response, the British developed Q-ships. These were merchantmen armed with concealed batteries of powerful guns. Under the rules of war at the time, when a U-boat surfaced alongside, it was supposed to wait until the ship's crew had taken to the lifeboats and then approach to take the ship's papers as evidence

ABOVE A British convoy on the Arctic route to the Russian ports of Archangel and Murmansk in World War II; ships grouped together under naval escort for protection against attacks by German aircraft and submarines.

OPPOSITE TOP A U9 German submarine became a powerful weapon during World War II.

BELOW In World War II, the German submarine threat reappeared in all its menace: The U-boat with its deadly cargo of torpedoes **(OPPOSITE)** became one of Nazi Germany's most powerful weapons.

that it was carrying war materials before sinking it. But the crew of a Q-ship would wait for the vital moment and then hoist the white ensign of a warship, drop the screens hiding the guns and open fire on the submarine. All that was needed was a direct hit through the pressure hull to cripple the U-boat, which would be left to sink.

Everything possible was done to keep the Q-ships a closely guarded secret. Their crews wore civilian clothes on and off duty, and many were given white feathers by civilians who considered them to be cowards avoiding military service.

But stories began to filter out from captured U-boat men and merchant seamen whose vessels had been rescued by the timely arrival of a Q-ship. It was only a matter of time before the Germans learned the truth, and the existence of the ships was confirmed by an agent named Jules Crawford Silber, who had overheard the rumors. As a result, the use of this desperate weapon rebounded on the British with catastrophic effect. Certain factions within German naval circles had long pushed for unrestricted submarine warfare, and now their hand was strengthened. It was decided that, in

future, U-boats would approach their prey submerged and sink the ships with torpedoes before escaping to safety.

These tactics changed the face of naval warfare. Merchant shipping losses climbed to unsustainable levels, and by 1917 Britain faced starvation. Only the reluctant adoption of a tactic from the Napoleonic Wars saved the situation. Merchant ships were grouped into convoys, which sailed under the protection of naval escorts, and losses dropped to a much less damaging level. Moreover, the deaths of American citizens caused by German

BACKGROUND AND OPPOSITE Norwegian survivors of the torpedoed liner *Athenia* sunk on the day the war began.

submarines proved to be another powerful factor in persuading the United States to join the Allied effort.

The perils of overcaution

Even when agents provide their controllers with important and worthwhile information, it may not be exploited properly, or it may be overtaken by events. The Germans were presented with an espionage coup on the very eve of World War I, which could have altered the outcome of that terrible conflict. The author of this achievement was a successful and long-time agent, an Austrian named Baron August Schluga, who had given the Prussian High Command the entire Austrian order of battle and profiles of the chief Austrian generals during the Austro-Prussian War of 1866.

Afterward, Schluga had moved to Paris, where he continued to supply secret information on the French to the Prussian military attaché. But for most of the following 45 years, he was left in place as a long-term investment. Finally, on the eve of the greatest war in history, a signal was sent to Schluga asking him for details of French plans and intentions. He responded with jewels in espionage terms—the entire mobilization plan for the French Army and, of far greater value, Plan 17, the strategic thinking of Marshal Joffre, France's military commander in chief. The German commanders could hardly believe their good fortune. In the end, however, they chose not to believe the information, on the grounds that it may have been foisted on Schluga by the French to mislead their opponents. A priceless strategic advantage had been thrown away through overcaution.

True or false? This was the question that constantly exercised the minds of those who directed the efforts of spies and spy networks during the two world wars of the 20th century. The established organizations of European espionage in Germany, Austria, Russia, France and Britain strove to uncover vital secrets from their adversaries during World War I. America joined the conflict in 1917, but had no espionage presence in Europe, where the decisive battles were being fought and where US troops were becoming engaged. However, in the even more widespread campaigns of World War II, the United States found that espionage was a vital weapon, the value of which could no longer be ignored. In tackling Nazi Germany through the efforts of the Office of Strategic Services (OSS), the USA was laying the foundations for the Central Intelligence Agency (CIA) and the National Security Agency (NSA), those giants of Cold War espionage.

BELOW Marshal Joseph Joffre, commander of the French Army at the outbreak of World War I and author of the vital Plan 17, betrayed to the Germans by an Austrian agent, Baron Schluga.

Since the very beginnings of espionage, every spy has had to lead a difficult and dangerous existence. In some cases, searching for the information needed by those to whom he or she owes allegiance brings with it the danger of exposure and arrest. In others, gaining access to the information is simple enough, but copying it or smuggling it out of an office, an institution or across a border produces the greatest peril. In almost every case, detection means the threat of interrogation, followed by execution or long years in prison and the likelihood of the spy being disowned by his or her own service. A vital channel of information may be cut off or misused by the enemy, with or without the cooperation of the agent involved, to feed false information back to those who used the agent in the first place.

With stakes of this magnitude, only the most highly motivated people tend to contemplate work as agents. They may be spurred by ideology, patriotism, hatred of a particular system or financial reward. No matter what drives them, the price of their survival lies in the skill with which they carry out their work, the quality of what the intelligence services call their tradecraft. This covers fundamental aspects of espionage, like the strength of their cover story, and details like the routines used for passing information and contacting their controllers, how they can tell if they are being followed and how to shake off their shadowers if they are.

Many spy novels have described the techniques used to watch for observers. If an agent changes direction without warning, on foot or at the wheel of a car, any potential shadowers will reveal themselves by following suit. Consequently, tailing a suspect can involve a whole team of shadowers, one handing over to another at any attempt by their target to evade them—by hailing a cab or leaping on (or off) a bus or subway train without warning. Sometimes an agent may change his or her image by dodging into hiding and reemerging without a coat before heading in a different direction, or striking up a conversation with a passerby.

Traditionally, spy networks protect themselves by grouping their agents into cells, so that if one is caught, the amount of damage to the entire network is limited. In cases where radio communications are involved, operators may be instructed to include a prearranged deliberate mistake in their transmissions to show that they are still able to operate freely. A perfectly correct message, or one with the wrong kind of mistake, would show that the agent had been captured and was operating under duress.

Passing on the information collected by an agent can be achieved in different ways. Often, an apparently casual meeting, prearranged by secret message or by leaving a sign on an agreed landmark, like a chalked mark on a streetlight, can be used to transfer a note or a roll of

film from agent to supporter. In some cases, "dead drops" are used. The agent simply leaves the material in an innocuous place from which a supporter picks it up after spotting the prearranged warning mark.

Over the years, much has been done to guard against the inherent risks involved in passing on information. At one time, spies tended to use invisible inks, chemicals that could not be seen when they dried, so that the message could be written between the lines of an innocent-looking letter, then revealed by warming the paper or brushing it with the right chemical. The earliest types of invisible ink, like sugar solution, milk and lemon juice, appear when heated, but copper sulphate reacts to brushing with a solution of sodium iodide.

More recently, the invention of the microdot, which allows a photograph of a page of information to be reduced to the size of a period on a printed page, has made it possible to hide a mass of information beneath a postage stamp, or as part of a typed letter, which can be sent through the mail in the ordinary way. On the other hand, the development of the compact cellular phone must make life a great deal easier for teams of shadowers trying to keep track of a potentially elusive subject.

The truth is that espionage remains a very vulnerable profession, especially in societies where rules and regulations tend to expose those who do not belong. For an agent to be truly successful, he or she must blend into the host society as perfectly as possible; even the most trivial of details may reveal them as impostors. In Britain during World War II, German spies were frequently Irishmen, who spoke English well enough, but their ignorance of wartime shortages and regulations caused them to be reported to the authorities. One asked for a drink at

ABOVE Espionage equipment. A fake flashlight with a flase cell used for carrying paper rolls of codes and information and a cigarette lighter with secret compartments.

OPPOSITE Radio transmitting and receiving equipment.

an inn during the afternoon, when everyone knew that the liquor laws made this impossible; another asked for a whiskey at a bar when this had been unobtainable for months. Strangers in quiet rural communities were immediately suspect, especially when wearing clothes with cut, material or makers' labels indicating overseas origins. To guard against this kind of detail giving agents away, Britain's Special Operations Executive insisted that, among other things, they had their British dental work remade in Continental style before they were dropped into Occupied Europe.

The Oslo Report

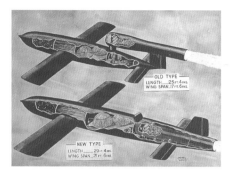

ABOVE The first of Hitler's revenge weapons: the original cruise missile in the shape of the V1 'Doodlebug' and the first ballistic missile, the V2 rocket (**BELOW**) being raised to its vertical firing position aboard a mobile launcher.

In November 1939, during the early days of World War II, Captain Hector Boyes, the British naval attaché in Oslo, found a note in the mailbox of his official residence, offering information on a wide range of German scientific and technical developments. To accept the offer, all the attaché had to do was arrange for the BBC to begin one of its German-language broadcasts with the words "Hallo, Hallo, hier ist London," instead of the usual single "Hallo." Afterward, a larger package appeared in the attaché's mailbox, containing seven pages of typewritten material and a glass tube, which turned out to be part of a proximity fuse for antiaircraft shells.

The text described the dive-bomber version of the Junkers 88 aircraft, work on radio-controlled glider bombs being carried out at Peenemünde and the development of early-warning radar systems and a precision guidance system to steer bombers accurately to their targets. It also referred to German naval advances in acoustic and magnetically triggered torpedoes. The entire package was passed to Professor R.V. Jones, wartime head of Air Ministry Intelligence, who asked for the opinion of Admiralty scientists. They were convinced that the whole package was a plant, designed to waste British Intelligence efforts on nonexistent targets, since they did not believe that any individual could know about so many different secret projects. Jones, having discovered that the proximity fuse was genuine and better than anything developed by the British, thought differently.

In the end, he kept the document beside him for the rest of the war, as a signpost to other areas of investigation, and stage by stage, most of its contents were revealed as being close to the truth. In conjunction with other information, it helped unravel German radar secrets and combat blind bombing guidance beams, radio controlled antishipping bombs and ultimately the V1 and V2 missiles.

But who was the unknown author of the Oslo Report? Who had had the breadth of knowledge and a wish to reach British Intelligence, and to do so in excellent German? Even after the war, Professor Jones had hoped that the writer would make himself known, but nothing was heard. Examination of the text showed that he had been most accurate when referring to electronics, and several candidates were suggested. One was a German engineer, Dr. Hans Kummerow, a member of the anti-Nazi resistance, who had been arrested and executed after the bomb plot on Hitler's life of July 20, 1944. Another was an Austrian named Paul Rosbaud, scientific advisor to the German Springer publishing group, who had been a prisoner of the British during World War I, and who had passed on information concerning German wartime nuclear experiments. But neither had written the Oslo Report.

The true identity of the author finally emerged after a long chain of coincidences, among them a meeting between Jones and a British engineer called Cobden Turner on the Queen Mary, when the former was returning from a conference in the USA. Much later, in the 1950s, it emerged that a prewar friend of Turner's had been the writer of the letters. This was Hans Ferdinand Meyer, who had been wounded in World War 1, studied at Karlsruhe and Heidelberg and worked for the Siemens electronics company. Worried that development of the secret weapons would give Nazi Germany victory, he decided that only by alerting the British could he help defeat the Nazis and restore Germany to normality. He had typed the letters on a borrowed machine in Oslo's Hotel Bristol while on a business trip to neutral Norway. Years later, he had been arrested for listening to BBC broadcasts and imprisoned in a succession of concentration camps. He survived the war and moved to the USA, where eventually he was appointed Professor of Signals Technique at Cornell University.

ABOVE German weapons being examined after the war. In the foreground an experimental piloted version of the V1 cruise missile, and in the background the deadly Hs293 radio controlled glider bomb.

BACKGROUND V2 rockets were fired against London and Antwerp causing terrible damage.

Richard Sorge and the Red Orchestra

When Hitler invaded Russia in June 1941, the Soviet leaders were desperate to know the intentions of Germany's ally in the east, Japan. With German divisions pushing deeper and deeper into the Russian defenses, huge Russian forces were tied down in Siberia, watching for a possible Japanese attack. But if the Japanese had other plans, these troops could be rushed to the fighting in the west, where their presence might tip the balance.

The Russians' master spy in Tokyo was an agent of mixed German-Russian parentage named Richard Sorge, who posed as a journalist with Nazi sympathies, so successfully that Josef Goebbels, Hitler's Minister for Propaganda, was a guest at a dinner celebrating Sorge's departure for Japan. His network included a young Japanese journalist, Hotsumi Ozaki, who had secret Marxist sympathies and who was a trusted advisor to a group led by Prince Konoye, which influenced Japanese foreign policy. Ozaki knew of Japanese intentions at the highest level, and could also argue for his and Sorge's point of view.

Although Sorge's network was starved of funds, with a total allowance of 1,000 dollars a month, which was cut repeatedly, he was able to warn Russia that the Germans would attack on June 20, 1941, but Stalin chose not to believe the information. German forces actually crossed the Russian frontier two days later.

At last, with the Russians retreating toward Moscow in the fall of 1941, Ozaki brought Sorge the information he needed.

TOP Richard Sorge enjoyed access to Nazi chiefs like Propaganda Minister Josef Goebbels **(ABOVE)**, enjoying a joke with his wife and Hitler at his office in Berlin, but was able to warn his Russian spymasters of the coming German invasion of their country in 1941 **(LEFT)**.

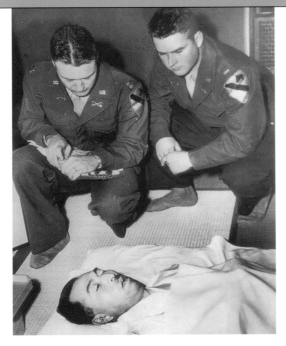

When Japan attacked her enemies at the beginning of December, she would strike south and east against the Americans and the British, not north and west against the Russians. This time, Stalin believed his master spy, and with the leading German tank crews in sight of the spires of the Kremlin, half the well-trained and well-equipped troops stationed along the Manchurian frontier were boarding westbound troop trains. By the time they arrived at the front, the terrible Russian winter had brought movement to a halt, but they were able to bolster the defenses when hostilities expanded in the spring thaw.

But within days of his final message, Sorge's network was uncovered by the Japanese police. All 35 of its members were put on trial, and Sorge and Ozaki were hanged.

However, Stalin had other spies at work in the west. The Red Orchestra was a prewar Soviet spy network operating first in Belgium, then in Holland and Germany itself, and also neutral Switzerland. Once Germany had attacked Russia, Stalin ordered the group to send more and more detailed information on the German armed forces.

For a year, the Red Orchestra passed on valuable intelligence. But on December 13, 1941, German radio direction finders in Brussels traced coded transmissions, each lasting for over five hours, to a group of three houses on the Rue des Attrebats. There they arrested three agents and seized equipment that included invisible ink, codebooks and false papers. Unfortunately for them, the chief of the entire European Red Orchestra organization, Leopold Trepper, arrived at the house while the search was in progress. His cover story of having rabbits to sell, in a city where meat was in short supply, was believed and he was released.

ABOVE Sorge's prime contact in Japan was a journalist, Hotsumi Ozaki working as advisor to Prince Konoye.

TOP Prince Konoye's body was found after he committed suicide following the Japanese defeat in 1945.

In June 1942, the Germans discovered another transmitter and the chief radio operator of a different part of the network, Johann Wenzel. His information led to the capture of the Red Orchestra and the arrest of Trepper. Because the spies did not have time to warn their controllers in Russia, the Germans were able to turn the network around and feed false information to the Soviets for the best part of a year. Eventually, however, the Russians became suspicious and started demanding more sensitive information, which the Germans could not supply. Finally, the transmissions ceased and the Red Orchestra lapsed into silence forever.

The French painter

B y 1943, it had become clear that soon the Allies would have to attempt a landing in Nazi-occupied Europe, and the Germans were rapidly fortifying the entire Channel coast with bunkers, gun emplacements, minefields and barbed wire. To give the landings the best chance of success, it was essential that the Allies discovered as much as possible about these defenses, but there was a limit to what could be obtained by airborne reconnaissance. What they really needed were the German engineers' plans, but clearly that was an impossibility.

However, on the morning of Thursday, May 7, a French decorator and resistance agent named Duchez spotted an official notice at Caen in Normandy, inviting estimates for the cost of repairs and refurbishments to the local offices of the Todt Organization, the German body responsible for building the defenses.

Duchez was a determined and resourceful man, who realized his opportunity. He had grown up in German-occupied Lorraine before World War I, and he spoke a little German. He decided to visit the Todt offices, where he found that the work involved wallpapering two upstairs rooms. He put in a quote at a third less than the true cost, and he was told to bring some wallpaper pattern books on the following day.

BELOW Detailed maps of the German defences along the Western side of the Cotentin peninsula in Normandy, with the aid of reports from agents in occupied France.

On the Friday, Duchez was discussing the work with Bauleiter Schnedderer, the head of the Todt's Caen office, when an assistant brought in a pile of maps. Among them, to the Frenchman's astonishment, was a master map showing all the fortifications planned for the Normandy coast. Concealing his excitement, he continued discussing the patterns, while the maps were left on a corner of the desk.

When the German was called into an adjacent office, Duchez made his move. It was far too dangerous simply to walk out of the building with the vital map, so he hid it behind a heavy wall mirror, hoping for a chance to spirit it away later. He left the other maps, trusting that the Germans would not notice the loss of one among so many. Schnedderer returned, they picked a pattern, and Duchez agreed to start on the following Monday.

After an anxious weekend, the courageous Duchez returned to the Todt offices. But when he asked to see Schnedderer, he found that the German had been transferred to St. Malo and replaced by Bauleiter Keller. Duchez asked Keller when he should start work on what had been Schnedderer's office, where the map was hidden. Keller replied that the contract didn't cover that particular room, but Duchez—thinking quickly—explained that he and Schnedderer had agreed that, in return for more work from the Todt Organization in the future, Duchez would include that in the price.

The Germans were delighted that they had struck such a good bargain. When Duchez said the work would take two days, the Germans promised to clear the room in preparation. The horrified Duchez insisted that this was not necessary, since he would supply enough drop cloths to cover everything. Finally, on the Wednesday evening, Duchez met his associates and the local resistance leader at the Café des Touristes with the map. But with an elderly German captain, known as Albert, sitting at a nearby table, he could say nothing. Then, while standing at the bar ordering a glass of calvados, he saw a Gestapo car driving into the square. He moved slowly to rejoin the others.

Eventually, it became clear that the Gestapo had passed, and Duchez was back at the bar, when Albert stood up to leave. Duchez reached for the German's greatcoat from the coat stand and helped him put it on. Albert left, and the others relaxed, asking Duchez what he would have done if the Gestapo had arrived to search the café. Duchez smiled and explained that, as a precaution, he had hidden the map in Albert's greatcoat pocket, and had retrieved it as the German left.

BACKGROUND Advance knowledge of these defences enabled the Allies to establish a beach-head and push inland to capture the key city of Caen.

BELOW Hitler meeting representatives of the Todt Organization, responsible for planning and building the fortifications along the French coast designed to repel any landings.

The Venlo Incident

TOP Major Richard Stevens and Captain Sigismund Payne Best **(ABOVE)** British Intelligence agents captured by German SS men in the Dutch border town of Venlo in the fall of 1939.

On November 8, 1939, Adolf Hitler spoke at the annual rally of old Nazi Party comrades in the Buergerbraukeller in Munich. A few minutes after he left, a bomb exploded in the cellar. Hitler suspected the hand of British Intelligence and ordered the SS foreign intelligence service to kidnap some of their officers. Fortunately, the SS had already organized a classic double-agent operation in Holland, involving an agent identified in their records as F479. The agent was a political refugee who had told the British that he was in touch with a large opposition group within the German armed forces, and was able to pass on a large amount of false information from his SS controllers.

During October, SS officer Walther Schellenberg had ordered F479 to arrange a meeting between him, posing as Hauptmann Schaemmel of the German Army Transport Office, and the agent's British contacts in Holland. The first contact, with Captain Sigismund Payne Best and Major Richard Stevens, together with a Dutch officer, Lieutenant Coppens, took place in the Dutch town of Zutphen on October 21, 1939. Discussions seemed promising and another meeting was arranged for October 30.

This time, when the Germans turned up at the rendezvous, a crossroads near Arnhem, the British failed to arrive. A Dutch police patrol arrested the Germans, questioned them and searched all their baggage before releasing them. Then the British arrived, claiming to have gone to the wrong crossroads, but Schellenberg suspected a setup, to check that the Germans were really who they said they were.

Schellenberg and his assistant were given a radio, codes and an emergency telephone number to contact in case of problems. The next meeting would be arranged by radio, and the British offered to lay on a flight to carry Schellenberg to London to meet other Secret Service chiefs. At this point, news of the Munich bomb changed the operation from a double-agent ploy to a full abduction. Schellenberg was called by SS Reichsführer Heinrich

Himmler in the middle of the night and ordered to snatch the British officers at their next meeting, due to take place at a café in the Dutch town of Venlo, close to the border with Germany, on the following day.

Schellenberg and his agents turned up at the café, but there was no sign of the British. They waited for an hour, and were about to leave when Best, Stevens and Coppens approached at speed in a large gray Buick. The car pulled into the car park behind the café and, at that moment, an SS car arrived with the snatch squad. Coppens opened fire, but was shot by one of the SS men. Another was about to shoot Schellenberg, who was in civilian clothes, but a member of the party recognized him and spoiled his comrade's aim. Meanwhile, the British officers and their driver were dragged into the German car, which roared off to the border. The two officers remained prisoners but survived the war, although with their capture went all hopes of a solid British espionage presence in Germany.

ABOVE Hitler delivering the speech at the annual Nazi Party rally in Munich before the detonation of the bomb which led him to suspect the involvement of British Intelligence, and order SS officer Walther Schellenberg **(BELOW)** to make contact.

BOTTOM Hitler greets Mussolini after surviving an attempt on his life on July 20, 1944.

The man in the Foreign Office

Although British efforts to establish a network of agents in Nazi Germany had been crippled at the start, the Americans were much more successful. The predecessor to both the OSS and the CIA was the innocuous-sounding Office of the Coordinator of Information, set up by President Roosevelt in 1941 and headed by a wealthy lawyer named William J. Donovan, who was better known by his nickname, "Wild Bill." In June 1942, the organization was renamed the OSS, and in November that year, Allen Dulles was sent by train through Vichy, France, to Berne, Switzerland, carrying a letter of credit for a million dollars, to set up a forward office as close to Germany as possible. His cover role was as special legal assistant to the American Minister to Switzerland, Leland Harrison, but word soon spread through the espionage underworld that the OSS was open for business.

Several would-be informers made contact with Dulles' office, but it was often found that the quality of their information was dubious when compared with data revealed by "Ultra" (see Chapter 5) and other forms of communications intelligence, and they were dispensed with. But in August 1943 came an exception to the rule, in the shape of Fritz Kolbe, a German courier who showed Dulles copies of German Foreign Office cables, promising to bring more the next time he could arrange a visit to Switzerland.

The agonizing question was whether the information was genuine or false and supplied deliberately by the Germans' own spy services—Kolbe had been turned down by the British for precisely that reason. Dulles was aware that there were two main dangers in treating the information as genuine. If he passed copies to London or Washington over the radio, it would give the German counterespionage service the opportunity of breaking

ABOVE Allen Dulles, wartime head of the American Office of Strategic Services (OSS) branch in Switzerland from November 1942.

the codes. It might also give the Germans the evidence they needed to have the OSS thrown out of Berne for compromising Swiss neutrality.

Nevertheless, the information was very convincing, and consistent with what was known through "Ultra" and other sources. Kolbe himself gave Dulles a wealth of personal information, which the Americans could check for themselves. On October 7, 1943, he delivered no fewer than 96 copies of messages, totaling more than 200 pages. He returned three more times with over 1,600 documents, many of them secret dispatches from German military attachés overseas. The information was priceless: it included details of Spanish dictator Franco's scheme for smuggling tungsten to Germany's steel industries in crates of oranges, and of a German radio station in neutral Ireland that reported on Allied shipping movements. Both operations were stopped after strong Allied protests.

Kolbe also revealed that there was a spy in the British Embassy in Ankara. The Albanian valet to the British ambassador to Turkey was able to steal documents from his employer's safe and pass them to the Germans. They included the top-secret plans for the invasion of Europe. Ironically, the Germans suspected that there was a British plant and refused to act on the information!

Other wartime OSS operations were also able to penetrate close to the heart of Hitler's Germany. Dulles ran eight networks in Occupied France that relayed vital details of German targets to be attacked in the run-up to the D-Day landings, and received information from German agents on the development of V1 and V2 weapons at Peenemünde. In the summer of 1944, the OSS moved onto the offensive, with 50 two-man teams, supplied by de Gaulle's intelligence service, being parachuted into France to report on German troop movements at the time of the invasion. Thirty-four teams of OSS agents were also dropped into Germany, under the authority of future CIA director William Casey. Four of the most successful were able to radio information back to specially equipped aircraft flying over the Reich at night, giving details of rail routes, troop movements and potential targets. One team, codenamed "Hammer," even managed to reach Berlin and was able to send back a message indicating German morale and defensive plans.

BELOW William Casey ran the program involving OSS agents in Germany and later directed the OSS's successor, the CIA.

Spies in Peacetime

ABOVE The directors of the Communist Cheka secret police organization set up its Moscow headquarters in 1918 to replace the Tsarist Ochrana, whose teams of undercover agents **(TOP RIGHT)** maintained surveillance on every level of Russian society.

pies have always faced particular perils in time of war, when national survival can depend on the timely information they are able to relay back to the intelligence services. The societies in which they live and work are likely to be most alert to anyone behaving suspiciously, or seen in inappropriate places or with unexplained contacts, particularly if they have sensitive jobs or access to sensitive information. But in peacetime, especially during periods of high international tension, like the Cold War, spies have been equally vital as a means of monitoring threats and occasionally reassuring countries that the danger of unexpected attack is not as great as their fears suggest. During the turbulent years of the 20th century, more and more countries came to realize that an active and competent espionage service was a necessary peacetime weapon.

Russia has a long tradition of espionage. In fact, the first secret police organization, the Tsarist Ochrana, was tasked with

LEFT Lenin realized only too well the need for a secret police organization to watch for internal and external enemies of his Russian Communist regime, and the Lubyanka building in Moscow **(ABOVE)** became both a detention and interrogation center.

spying on the Tsar's own subjects, to watch for signs of dissent or revolution, and followed a tradition first established by the paranoid Ivan the Terrible in the 16th century. Not only did it strive to place agents in all the dissident movements, but also it used agents provocateurs, who would lure their colleagues into undertaking unwise actions for which then they could be arrested and exiled to Siberia.

Following the October Revolution of 1917, which brought the Bolsheviks to power, Lenin's regime simply replaced the Ochrana with the Cheka, or the Extraordinary Commission for Combating Counterrevolution and Sabotage, the precursor of the KGB. As the fledgling Communist state faced threats

from outside as well as from within, this resulted in a major Russian espionage effort against other countries, an offensive that continues to the present day.

During that time, the nature of the external threat, real or imaginary, and the kind of information needed changed completely. Between the world wars, Russian espionage was concerned mainly with undermining the various opposition groups that had taken refuge overseas, and many agents were recruited from the Communist parties in those target nations. After World War II, however, the terrible death toll that had resulted from the German invasion made the Soviet leadership deeply worried about the possibility of another invasion from the west.

Chasing the atomic secrets

When the atomic bombs were dropped over Hiroshima and Nagasaki in 1945, forcing the Japanese to surrender, the Russians realized only too well how vulnerable they were, despite their vast armed forces and huge territory. They spared no effort in ferreting out the secrets of this awesome new weapon, so they too could enjoy the power that possession of nuclear weapons brought. Their success, to the point where the first Russian atomic bombs were tested in 1949, spurred them on to still greater efforts to wrest knowledge and information from the West, and triggered alarm bells that induced the West's intelligence services to redouble their efforts within the Communist bloc.

In America, President Truman had closed down the OSS in 1945, on the grounds that it had no place in the new peacetime world. Only when evidence of increased Russian belligerence showed that this was premature did he set up a new organization called the Central Intelligence Agency, under the control of the National Security Council (NSC) and incorporating links between the military intelligence arms of the three services. Based at Langley in Virginia, the CIA soon gained a formidable reputation for the quality of its intelligence gathering, despite the fact that the USA had had little tradition of espionage during the long period between the War of Independence and the eve of World War II.

The Gehlen Bureau

In some areas, the West was able to hire professionals with close and recent links to intelligence gathering. One of the foremost experts on the Soviets and their military capabilities, Reinhard Gehlen, was an officer in Hitler's Wehrmacht during World War II. He developed the wartime Foreign Armies, East section of German military intelligence, concentrating on their Eastern Front opponents and building a strong network of agents, with detailed knowledge of Russian strengths, weapons and capabilities.

As the war drew to a close, Gehlen made careful preparations. A lieutenant general, he ordered microfilm copies to be made of all his records and had the originals destroyed. The copies were hidden in the mountains of Bavaria, ahead of the advancing American Army. When Gehlen surrendered, he offered his captors vast quantities of information on the Soviets, and soon he was taken on by his new masters. In 1955, he switched employers again, to become director of the intelligence service for the new Federal Republic of West Germany, and during these years he produced enormously valuable information on Communist East Germany. Among his agents was a member of the East German Cabinet, while another worked for East German Intelligence.

One of Gehlen's agents managed to deliver a copy of the secret speech made by the Soviet leader Nikita Khrushchev to the Party Congress in Moscow in 1956, denouncing the legacy of Stalin; another predicted the Russian crackdown in Czechoslovakia in 1968, when the Soviet leadership became alarmed by the liberal reforms of the Prague leaders. But even Gehlen had his failures. His agents gave him no advance warning of the closing of the East-West border in Berlin in 1961 and the building of the Berlin Wall. Later that year, his reputation suffered further when it was learned that his own chief of counterintelligence, Heinz Felfe, had been a spy for the Russians for more than 10 years!

TOP Gehlen's chief of counter-intelligence Heinz Felfe was revealed to be a Russian double agent.

BACKGROUND The building of the Berlin wall in 1961 was to divide the city for 30 years and make it the European capital of espionage.

BELOW RIGHT Hitler's Chief of Intelligence on the Eastern Front, General Reinhard Gehlen built a new post-war career working for the West Germans.

BACKGROUND Tanks near the wall of the Friedrichstrasse checkpoint of the Berlin Wall impose East German police checks on American officials.

RIGHT The early structure crosses the middle of a street as the Berlin wall goes up.

TESTIMONIANZA FORNITA

Identità: Carta d'identità n° I3I rilasciata dal Comune di Termeno
l'II-VI-I948
(Documenti personali presentati)

Emigrazione: Permesso di libero sbarco n° exp.23I489/48
(Indicare se avverrà tramite un Comitato responsabile. Designazione dell'Autorità. Num. di registrazione)
Partenza col Piroscafo ANNA "C" nella prima metà di Giugno

o privatamente (indicare promesse di visto ottenute):

CONNOTATI

Capelli: castani
Occhi: celesti
Naso: regolare
Segni particolari:

Impronta digitale
(pollice destro)

Visto per l'autenticità delle dichiarazioni, fotografia, firma e impronta digitale del Sig. Klement Riccardo
Firma e timbro dell'Autorità: P. Domohig Edoardo
Luogo e data: Genova I/6/I950
(pregasi apporre il timbro anche sulla fotografia)

Carta 10.100 bis N. I00940 Validità un anno
Concessa a Genova il I/6/I950
Consegnata a " il " " "

Israeli Intelligence

While the East-West espionage struggle was largely devoted to three things—the opposing bloc's strengths, intentions and nuclear secrets—other countries developed intelligence services with different priorities. One of the best was, and is, Mossad, the Israeli external espionage organization, which has much more specific objectives, mainly closer to home. It includes a Special Operations Department, which carries out actions against those thought to be a threat to the state, and to punish its enemies. In 1960, Mossad agents went to Argentina to track down and seize Adolf Eichmann, the Nazi official who had organized the transportation of European Jews to Hitler's death camps during World War II.

Other Mossad operations include the hunt for the perpetrators of the murder of the Israeli athletes at the 1972 Olympic Games in Munich. Those found to be associated with the assassinations were themselves assassinated: one by a car bomb, another by a bullet and a third by explosives activated by his telephone. Four years later, when an Israeli aircraft was hijacked to Entebbe in Uganda, the Israelis mounted a long-distance raid that liberated the captives and took them safely home.

LEFT AND FAR LEFT Nazi official Adolf Eichmann lived after the war in Argentina on forged identity papers **(TOP LEFT)** and was seized by Israeli agents in 1960 to be put on trial for his involvement in the extermination of European Jews **(BOTTOM)**.

OPPOSITE Poster for the 1972 Munich Olympics where Israeli athletes were seized by Arab terrorists and killed in the subsequent shoot-out, prompting a long and successful campaign by Israeli intelligence to track down those responsible.

München  1972

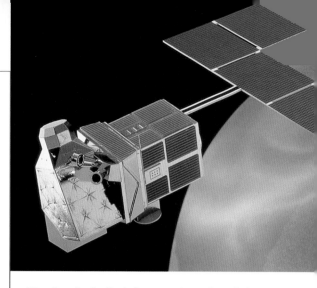

Mutual Assured Destruction

For many of the postwar years, an uneasy peace between East and West was maintained by the balance of terror between the superpowers. Because of the awesome power of nuclear weapons, the only sure means of living with the fear of an overwhelming attack was by convincing the opposition that any strike would bring massive retaliation. This was the doctrine of MAD—Mutual Assured Destruction—which partially reversed the priorities of espionage in this particular area.

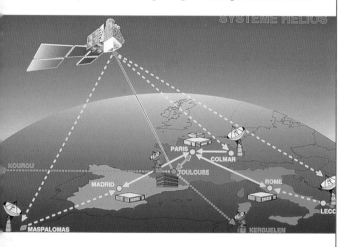

For deterrence to work, it was essential that opponents were convinced of two things: the power of the nuclear forces available and the determination to use them if attacked.

On the other hand, if a nation's spies could find out any details of the locations and defenses of the nuclear forces belonging to an adversary, this might make the threat less overwhelming and tilt the balance in that country's favor. Hence the importance of intelligence, to gain an advantage over the opposition, and counterintelligence to ensure that the opposition was not able to win an advantage in turn.

Factors like these occupied the espionage services of the countries of NATO and the Warsaw Pact right through to the 1990s, when the American decision to develop the "Star Wars" antimissile defense systems forced the Soviets to engage in an arms race they clearly could not win. With surprising suddenness, the East-West balance collapsed like the proverbial house of cards. The Soviet frontiers receded from the River Elbe far to the east, and the Warsaw Pact satellites jumped ship, to adopt neutral status or to sign up with NATO, as the only credible military-alliance game in town.

Not that this diminished the popularity of espionage; quite the contrary. Without the clear-cut priorities of Cold War rivalry, however, tomorrow's threats seem much more hazy and indefinite. Weapons of mass destruction, chemical and biological as well as nuclear, are now more easily hidden and more easily delivered than ever before, and the hazards of terrorism and religious fundamentalism join those of ideology and power politics as justifications for keeping the closest of eyes on a country's allies, neighbors, trading partners and competitors as well as opponents. Much of that effort has been directed into airborne and satellite reconnaissance, electronic intelligence and image intelligence, as will be explained in later chapters. But examples show all too clearly that the well-placed and well-motivated human agent is still a powerful player in a paranoid world.

ABOVE LEFT Diagram of the range and capability of Helios **(ABOVE)**, the first European defense satellite.

OPPOSITE Characteristic nuclear explosion mushroom cloud that has been an image of terror for generations.

Spies in the slide to war: Europe 1914

For much of the time, spies work in secrecy and isolation, never knowing whether the information they risk their lives to find will be vital to their masters, or merely provide one more small piece in the huge jigsaw puzzle of the overall intelligence picture. Yet there have been occasions when the most minor of observations by agents lucky enough to be in the right place at the right time have changed history. In many cases, they have succeeded in helping the world to avoid the catastrophe of war. But on one occasion, the dedicated and selfless sacrifice of spies of several different countries conspired to tip the world over the edge, into the bloody cataclysm of the "war to end wars" in the late summer of 1914.

By beginning of the 20th century, war was no longer a matter for small, highly trained, professional forces, a military elite that fought its battles without involving the rest of the population. Ever since the wars of the French Revolution, fighting had involved whole nations, citizen armies that harnessed the entire strength of a country's manpower, wealth, weapons and resources. And the process of mobilizing these huge armies was a difficult and dangerous business. Delayed too long, it could leave even a powerful country at the mercy of an enemy that had proved to be faster out of the starting gate in a crisis where diplomacy had given way to battle.

In many ways, Europe in the early 20th century was a war waiting to happen. Imperial Germany's anger at having been left out of the empire building process, which had occupied most of its neighbors during the preceding 100 years, was finding a new objective in the Kaiser's demands for "a place in the sun" for his country, a new status to match its military and economic power. Already the British were worried about German plans to build a battle fleet that could challenge the world dominance enjoyed by the Royal Navy since Trafalgar, while the French were thirsting for an opportunity to avenge their defeat of 1871 and the loss of the provinces of Alsace and Lorraine. The Austro-Hungarian

Empire was anxious to punish Serbia for the assassination of the Austrian Archduke Franz Ferdinand in Sarajevo on June 28, 1914, by factions protesting at Austria's seizure of Bosnia six years before. A network of treaties and alliances, intended to make war less likely, bound Germany to Austria, Russia to Serbia, and France to Russia.

This deadly, but still stable, situation was finally upset on July 23, 1914, when the Austrians delivered an ultimatum to Serbia, threatening invasion if the Serbs did not give way on all the 10 points it contained by 6 o'clock on July 25. The Serbs sent their reply 10 minutes before the deadline, agreeing to all the points save one. Ten minutes later, the Austrians broke off diplomatic relations, and Austrian troops moved up to the border. After two days, they crossed it, and the invasion had begun.

Despite their belligerence, the Austrians were desperate to know what was happening in Russia, Serbia's most powerful ally. They contacted the chief of the German military intelligence service, Walter Nicolai, who called his team to full readiness. Officers in districts along the borders with Russia and France sent volunteer agents across each frontier, posing as businessmen and tourists, to watch for signs of mobilization. The German

TOP Archduke Franz Ferdinand, son of the Austro-Hungarian Emperor Franz Josef, assassinated in Sarajevo on June 14, 1914 by Gavrilo Princip **(RIGHT)**, being seized after firing the fatal shots.

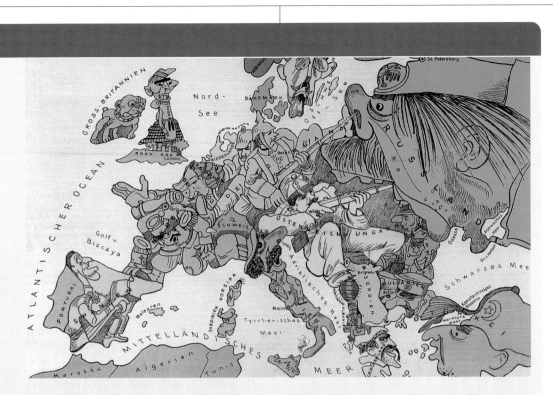

wireless monitoring service had picked up a long exchange of coded traffic between the French transmitter on the Eiffel Tower and a Russian station on July 24, and the next day a German attaché at the court of the Tsar reported that Russian troops at camps near St. Petersburg had been ordered back to their regiments.

More and more information was revealed by the agents. Russian border guards were being placed on full alert, and Russian troop trains had been seen heading toward the border, while empty freight trains were moving in the opposite direction, ready to pick up more loads. Martial law had been declared in several Russian border districts. At a top-level meeting on July 28, the Germans concluded that the Russians had begun their premobilization routines, but that France remained quiet. Two days later, the Germans noted that there was no evidence of either Russian or French reservists being called up, and entertained cautious hopes that the situation remained stable.

The following day, July 29, reports arrived from within Russia that partial mobilization was under way. Late that evening, spies indicated that the order had gone out from St. Petersburg for full mobilization to begin. At first, the German General Staff refused to believe these reports, but on the next day, agents reported seeing red posters announcing full mobilization in all the Russian border districts. On July 30, Germany and Austria sent a joint

ABOVE A cartoon map of 1914 showing Europe as a chaos of colliding nationalisms.

ultimatum to the Tsar, demanding that mobilization be halted by noon on August 1. The Germans also sent a demand to France, insisting that she stay neutral if Germany, Austria and Russia went to war.

No reply came from the Russians. The German ambassador in St. Petersburg was ordered to deliver a declaration of war by 5 o'clock that afternoon. At the same time, the Germans ordered full mobilization, although some preparations had already begun. War fever seized Berlin, and several real or suspected Russian spies were beaten to death by large and bloodthirsty crowds.

Unfortunately, the huge, supremely detailed and efficient German mobilization plan, involving 11,000 trains traveling to a Europe-wide schedule just 10 minutes apart, made no distinction between preparations on the Russian and French frontiers. Equally unfortunately, it depended upon the seizure of a vital rail junction just across the frontier in neutral Luxembourg, by 7 o'clock in the evening of August 3. And most unfortunately of all, it required the taking and crossing of neutral Belgium, which would bring the British into the war. At this stage, the spies had done their work, handing over the crisis to the politicians and, ultimately, the soldiers—all that remained was the fighting.

The atom bomb spies

ABOVE Anatoly Yakovlev a clerk at the Soviet Embassy who received Gold's information.

BELOW The Russian spy Harry Gold, after being arrested by two United States Deputy Marshals.

On the morning of Sunday, June 3, 1945, Harry Gold, a chemist from Philadelphia, called at the home of David and Ruth Greenglass at 209 North High Street, Albuquerque, New Mexico. He identified himself by producing the torn half of a cardboard panel from a carton of raspberry-flavored Jello, which matched the half produced by David Greenglass. In return for an envelope containing 500 dollars in cash, Greenglass handed over some handwritten notes and sketches, which Gold placed in a large manila envelope that matched another he carried in a folded newspaper.

The other envelope contained the fruits of a meeting he had had the day before in Santa Fe. He had arrived there at 2:30 pm and had bought a tourist map from a local museum so that he could find his way to the Castillo Street Bridge without having to ask directions. There, at 4 o'clock, he met a German-born British scientist, Dr. Klaus Fuchs, who offered Gold a ride. After a short drive, during which Fuchs explained the nature of his work, the scientist handed Gold a sheaf of notes, which went into the manila envelope.

After leaving the Greenglasses, Gold traveled by train to New York, arriving late Tuesday evening. Then he took a roundabout route to the boundary between Brooklyn and the neighboring district of Queens. Just before 10 o'clock, he met a clerk from the Russian Embassy named Anatoly Yakovlev. They spoke briefly and exchanged newspapers before walking away in opposite directions. The newspaper handed to Gold was empty, but Yakovlev left with enough information to give his country a head start in manufacturing an atomic bomb of its own, saving years of expensive development work.

The development of the nuclear bomb, under conditions of the highest security in the U.S.A. during World War II, had changed the face of warfare. The British and Americans had calculated that they could face as many as three million casualties if forced to invade a fanatically defended Japan in 1945 to end the war. Simply detonating two atomic weapons over Hiroshima and Nagasaki achieved the same result, with no Allied casualties at all.

Stalin was horrified. How could the Soviet Union, still recovering from the war, hope to catch up the West without decades of research and development? The only viable alternative was to steal it.

Fortunately, there was a large band of Communist sympathizers with access to those priceless secrets. Russian espionage had begun in the USA while the bomb was still under development. Yakovlev's network included Julius and Ethel Rosenberg, both of whom had been active spies during the war, when Julius worked for the US Signal Corps. However, their breakthrough came in 1944 when Ethel's brother, David Greenglass, was assigned to the Manhattan Project, the cover name for the development of the atomic bomb. In return for cash, he passed on vital technical details, including the method for casting the molds for the interior contours of the bomb.

Klaus Fuchs was also part of the network. He had passed information to the Russians while carrying out nuclear research at Birmingham University in England. Transferred to the Manhattan Project in 1943, he was ordered to contact Harry Gold, whereupon he started supplying details of American work on the bomb. Alan Nunn May was a British physicist working on nuclear research in Canada. On trips to the USA, he was given detailed technical briefings on the development of the bomb, which he passed on to a different Soviet spy network.

The reality of the achievement of this group of spies was finally revealed on September 23, 1949, when evidence emerged of a test explosion deep in the Soviet heartland. The West's immense lead in nuclear development had been stolen by a group of dedicated individuals, and given to a one-time ally, which now was demonstrating increasing, and worrying, hostility.

Retribution was not long delayed, however. When cipher clerk Igor Gouzenko defected from the Soviet Embassy in Ottawa in September 1945, the information he revealed indicated that Nunn May was a spy, and the physicist was arrested, tried and imprisoned. Gouzenko told the Canadians that the Russians had been given enormous amounts of information on the atomic bomb project by other spy networks, and the FBI was alerted. After detailed investigations, they passed on their concerns to the British, who became suspicious of Klaus Fuchs. He had returned to England in 1946 and was working for the UK Atomic Energy Research Establishment at Harwell, Oxfordshire.

Fuchs was arrested and questioned, and he revealed information that uncovered a trail to Gold, to Greenglass and, ultimately, to the Rosenbergs. All of them were put on trial. Gold, Fuchs and Greenglass were given prison sentences, but the Rosenbergs, who were portrayed as the ringleaders in the conspiracy, were sentenced to death. Despite a vast international outcry to spare their lives, they were executed in 1953.

ABOVE Senior members of the staff of the UK Atomic Energy Research Establishment including Klaus Fuchs **(FAR LEFT)**.

BELOW David Greenglass, shown here escorted by US Deputy Marshal, who passed on nuclear bomb information to the Russians for cash.

BACKGROUND View of Nagasaki from the bomb site as the atomic bomb that levelled the city was flown over it.

The Cambridge Group

ABOVE, LEFT TO RIGHT The members of the infamous Cambridge Group included Guy Burgess, Kim Philby, Donald Maclean and John Cairncross.

FAR RIGHT Anthony Blunt, leader of the Cambridge group while a Fellow of Trinity College. **(BELOW)**.

For the Russian espionage service, the 1930s were golden years for recruiting dedicated would-be agents in the West. In England, they reaped a rich crop of lifelong spies, in the shape of a quintet of Communist sympathizers at Cambridge University, led by a Fellow of Trinity College, Anthony Blunt. The group included four undergraduates: Donald Maclean, Guy Burgess, John Cairncross and Kim Philby. Although there was little that they could do at the time, the Soviets were more than prepared to wait until their protégés gained positions of influence, with access to important information.

Philby worked as a journalist after leaving Cambridge, taking care to conceal his Communist allegiance. He covered the Spanish Civil War from the fascist side and, following the outbreak of World War II, reported on the operations of the British Army in France. At the time, Guy Burgess was his courier, but it was not until after the evacuation at Dunkirk that Philby joined Burgess in working for MI6, finally moving to the Russian Department in 1944. Blunt, meanwhile, had transferred from the army to the counterintelligence service, MI5, while Maclean was working at the Foreign Office with Cairncross, who later moved to the Treasury.

Their Soviet masters had already seen a handsome return on their investment, but the value of their stock was to rise, before eventually it fell. In 1949, Philby was sent to Washington as liaison officer to the CIA, and Burgess was appointed Second Secretary at the British Embassy there, although later he was sent back to London with a heavy drinking problem. Maclean had also served at the Washington Embassy, but by 1949, with a drinking problem of his own, he had been appointed to the American section of the Foreign Office, which enabled him to supply his Soviet masters with secrets of Anglo-American plans for nuclear research.

However, the situation was too good to last for long. In 1951, a code-breaking success by the West revealed that someone was leaking nuclear secrets, and suspicion fell on Maclean. The information reached Philby, who tried to deflect suspicion on to someone else, but he learned that Maclean's interrogation would begin on Monday, May 28. It was time for the Russians to act. On the evening of Friday, May 25, Burgess called at Maclean's house, just as he was about to sit down to a birthday dinner, and the two men left on a ferry to France. There they received false papers and traveled on to Vienna, then occupied by the four Allied powers, where they crossed into the Russian sector and headed for Moscow.

Notes found in Burgess's flat implicated Cairncross, who denied that he was a spy, but admitted to having passed some information to the Russians, which led to him being forced to resign from the Treasury. Philby's warnings to the wanted men placed him under suspicion too, and he was recalled to London for detailed questioning. He managed to deflect most of the allegations, to the point where he was allowed to remain with MI6. In 1956, he was posted to Beirut, where he remained for seven years until, worried that he might be summoned to London to face fresh allegations, he finally defected.

Although he had been out of the espionage loop since the end of the war, Blunt had helped arrange the escape of Burgess and Maclean. In 1964, he was secretly accused of having been in the Cambridge Group, and he admitted his guilt to his interrogators. Because he revealed much of the Soviet espionage setup in the UK, he was left alone, but in 1979, his treachery became public knowledge and he was stripped of his honors, including the knighthood he had been given for his services to the Royal art collections. Finally, the five members of the Cambridge Group had been unmasked.

BELOW Guy Burgess first brought suspicion on the Cambridge Group while working at the British Embassy in Washington.

The Chancellor and the spy

ABOVE Günther Gulliaume with his wife Christel also ended the political career of West German Chancellor Willy Brandt seen with Guillaume, **(OPPOSITE)**.

BELOW Minutes of meetings like this one involving US President Richard M Nixon were passed on to the East Germans by Günther Guillaume.

Throughout the Cold War years, especially before the Berlin Wall made escape more difficult, political refugees from East Germany fled to the Federal Republic of West Germany, where they were granted automatic citizenship. Many of them were genuine opponents of the Communist regime, but some were planted deliberately on West German soil as potential agents for the East. One of the most successful was Günther Guillaume, who crossed the border in 1956 with his wife, Christel. The couple settled in Frankfurt, and a year later Günther joined the Social Democratic Party.

So solid were his right-of-center views, and so diligent his work, that by the 1960s Guillaume was gaining power and influence within the Party, although he had access to little or no useful information for his masters. However, they were content to wait, and in 1970 their patience was rewarded, when he was appointed to the office of Federal Chancellor Willy Brandt. Only one person opposed his appointment, General Reinhard Gehlen's successor as head of West German Intelligence, General Gerhard Wessel, who was suspicious of Guillaume's background in the East.

Nevertheless, Horst Ehmke, head of the Chancellor's office, overruled Wessel and the appointment was made. As personal assistant to Brandt, Guillaume had access to priceless information: from biographies of the influential members of the party to the minutes of meetings with the West's leaders, like Nixon and Kissinger; and from reports of discussions among members of NATO to details of the West German attempts to counter spies from the East! Guillaume proved particularly useful when treaties were being negotiated between the two Germanies. Since he was able to reveal the West's bargaining position to the East Germans at every stage, his masters invariably came off best in all the subsequent arguments.

Even when he was unmasked, Guillaume proved valuable to those who had sent him across the border. He was arrested on April 24, 1974, and when the truth was made public, the charismatic and highly pro-West Chancellor had no option but to resign. Guillaume was released and returned to East Germany in 1981, to receive honors and promotion for his invaluable service.

The tourist spies—
Operation Redskin

Not all spies are trained agents. Some are ordinary travelers making entirely legitimate visits, for business or pleasure, to the target country, and who are briefed by intelligence experts on exactly what to look for on their travels. This was the substance of the CIA's carefully organized and very productive Operation Redskin, which made the maximum use of American visitors to the Soviet Union during the Cold War.

For years, military attachés appointed to embassies in both East and West had used every opportunity to note significant information on their legitimate travels around the countries to which they were assigned. For example, in June 1953, the US, British and Canadian air attachés in Moscow were able to visit a military airfield at Ramenskoye, on the outskirts of Moscow. There they became the first Westerners to see the Tupolev TU-4 long-range bomber, and the American attaché was able to photograph the aircraft. In March that same year, an American military attaché was able to carry a wire recorder on a road trip to the airport, which picked up the frequencies of radar signals in the area.

From then on, the USA and UK operated a coordinated program to make use of every official visit to Russian territory, by providing diplomats and embassy staff with travel folders specifying the kind of details to look for. Finally, the CIA built on

BELOW British and American visitors to Russia who flew on internal flights operated by the Russian airline Aeroflot were briefed to look for specific details on their trips as part of Operation Redskin.

these foundations to set up Operation Redskin, a means of providing nondiplomatic visitors from the West with a list of things to watch for on the other side of the Iron Curtain.

This was espionage for the masses, and without risk. The participants had no need to depart from their routes, or to make contact with people within the Soviet bloc. Instead, all they had to do was make notes of, for example, the color of the smoke from a particular factory chimney, or the registration letters of the airplane on which they flew. Whether they were tourists, businessmen, academics or journalists, the details they were asked for were added to a mass of information that revealed more and more about Soviet resources, equipment and production rates and methods.

As travel restrictions for tourists were eased within Russia and her satellites, progressively more places were added to tourist itineraries. Several of these were reached by propeller airplanes flying at relatively low altitude, allowing photographs to be taken of the ground below. Overall, the Redskin program produced details of surface-to-air missile sites and more than a dozen tests of these missiles in progress, together with valuable information on ICBM sites and production facilities.

Other travelers brought back details of early Russian nuclear submarines and missile-armed destroyers. But the traffic was not all one way. While British forces were still based at Singapore, air traffic controllers noted that a high proportion of civilian airliners overshot the runway and had to orbit the airport before making another attempt to land, which invariably was successful. When they checked, they found that the specified overshoot pattern took the aircraft over a secret radar site, and almost all the aircraft involved were from Warsaw Pact countries!

ABOVE The Russian Tupolev TU-4 long-range bomber, closely based on the American Boeing B-29 and B-50 Superfortresses.

ABOVE Diagram of a TU-4 model undergoing a shadow test, where the dark areas shown when illuminated by a beam of light revealed its most vulnerable areas.

The Penkovskiy case

ABOVE Presented with evidence which included spy cameras and rolls of film **(EXTREME RIGHT)** and coded writings on postcards **(BACKGROUND RIGHT)**, the judges at the Penkovskiy and Wynne trials found both men guilty of espionage.

TOP AND ABOVE RIGHT BACKGROUND Oleg Penkovskiy hears his death sentence announced in court.

Not all the successful spies were operating against the West. Several Russians, disillusioned with the actions of their masters, with the Communist political system and the restrictions it imposed on both thought and freedom, or the treatment it meted out to dissidents, worked actively for the West to provide vital information in the face of the most incredible risks. One of the most successful was Colonel Oleg Penkovskiy, who had gained a distinguished service record in the wartime Red Army fighting against the Germans, and who had been selected to serve in the GRU military intelligence organization during the 1950s. By the end of that decade, he had become deeply worried that Khrushchev's leadership might trigger a world war, and he decided that the only way he could work to prevent this was by collaborating with the West, to alert them to the threat, and give them the information that might help them to defuse it.

In April 1961, Penkovskiy approached Greville Wynne, a British businessman who was in Moscow to arrange the facilities for a Russian trade delegation's visit to the UK, which would be led by Penkovskiy. Wynne had actually been involved in intelligence work during and after the war, and he agreed to communicate Penkovskiy's approach to the right people in London. When the Russian delegation arrived in the UK, Penkovskiy was able to meet representatives of the British and American intelligence agencies, who checked his credentials and decided he was not a deliberate plant working for Moscow. At every possible opportunity during the delegation's visit to Britain, he met the Western controllers and told them all he knew. Communications were set up, he was given a miniature camera and a radio, and it was agreed that he would send back more information after his return to Moscow on May 6, 1961.

When Wynne next visited Moscow, on May 27, Penkovskiy handed him 20 rolls of film, covering everything from operating manuals for missiles to Russian intelligence documents. On further visits to London and Paris, in July and September, Penkovskiy supplied more information and took back material provided by his Western contacts, which he could present to his Soviet masters as evidence of his intelligence work for them in the West. He was given another contact in Moscow, the wife of an attaché at the British Embassy, to whom he passed exposed film in a box of candies for her children.

After returning from Paris, Penkovskiy continued passing on material to different contacts, but at a meeting with the attaché's wife on January 5, 1962, he spotted someone watching them, and noted the number of what seemed to be an official car. Their next meeting, a week later, went as planned, but on the following week the car returned, so Penkovskiy switched to other methods. These were mainly "dead drops," like the one at 5/6 Pushkin Street, where to the right of the front door was a radiator, behind which messages could be placed. To signal his contacts that a message was waiting, he had to make a black mark on streetlight number 35 on Kutuzov Prospect, or call two prearranged telephone numbers and let the phone ring for a set number of times before replacing the receiver.

However, it was clear that the KGB were growing increasingly suspicious of his contacts with the West, and when Greville Wynne arrived in Moscow on July 2, 1962, he found that Penkovskiy was convinced he was being watched. They arranged to meet again at the Peking Restaurant three days later, at 9 o'clock, but in the meantime Penkovskiy had discovered that the KGB were suspicious of Wynne. When Penkovskiy arrived at the restaurant, his experienced eye spotted a whole squad of KGB agents. He left the restaurant immediately and waited outside for Wynne.

The two men risked only a few brief words. Wynne was due to leave Moscow on the following morning, and Penkovskiy promised to see him off at the airport. Then he went back to his office and made a formal complaint against the KGB surveillance, since he had reported Wynne as a contact who could be useful to the GRU as a cover story to explain their meetings. The KGB apologized. By now, Wynne was worried enough to go to the airport at 5:30 in the morning, even though he was booked on a London flight leaving in the afternoon. Penkovskiy arrived at 6:15 and changed Wynne's ticket for the first flight to the West, the 9 am departure for Copenhagen. The plane took off, with Wynne aboard, and it seemed that the KGB had been outwitted.

In the end, though, they had their revenge. After passing onto the West the details of Russian missiles, which enabled American reconnaissance to identify the missile types being installed in Cuba during the ensuing missile crisis, Penkovskiy was arrested by the KGB on October 22, 1962. Wynne remained at liberty for another 11 days before he too was arrested while setting up a trade exhibition in Budapest, Hungary. The two men next met in the dreaded Lubyanka prison in Moscow, and they were put on trial in the following year. Both were found guilty of espionage. Penkovskiy was shot, while Wynne was imprisoned. A year later, he was exchanged for KGB Colonel Konon Molody, who had been captured in the UK under his cover name of Gordon Lonsdale.

ABOVE A year after Oleg Penkovskiy was sentenced to death, Greville Wynne (shown with his wife) is released in exchange for Soviet agent Gordon Lonsdale.

Codes and Ciphers

From the very earliest times, spies have depended on a variety of methods for concealing the text of messages they carried, or left for others to pick up, from those who might intercept them. The rise of cryptology, which employed different systems for representing the letters of a message with other letters, so that it was completely unintelligible to anyone who did not know the system used, was eventually matched by developments in the art of decrypting. It was the Arabs, some 700 years ago, who first laid down the principle of looking for the least used and most common letters in a message, and relating these to the least used and most common letters in normal language. They reinforced these clues by setting down the least used and most common combinations of letters in normal writing.

By the 16th century, diplomats in Renaissance Europe commonly employed several methods of encryption to render their dispatches and notes unreadable by outsiders, and the espionage services maintained by most

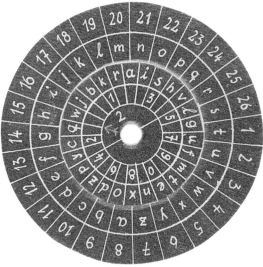

ABOVE Leon Battista Alberti, Florentine inventor of the cipher wheel.

RIGHT Early cipher wheel that uses revolving disks that can be aligned variously to encipher and decipher text.

monarchs tried every method they could find to decipher them. Some encryptors went so far as to employ elaborate codes, rather than ciphers: instead of changing the letters that made up a message, they relied on groups of letters or numbers, or even both, to represent names of countries, individuals and commonly used words. In a really comprehensive code, where almost every word was represented by one of these groups, it was virtually impossible to break the system and reveal the original text. However, given enough traffic in a particular code, much could be revealed by the content of successive messages.

ABOVE Rebel cipher encoding device captured at Mobile, Alabama during the Civil War in 1865.

The cipher wheel

To simplify the task of enciphering and deciphering a message by those entitled to do so, spies and diplomats soon began using mechanical devices. One of the earliest was the cipher wheel, invented by an Italian cryptologist named Leon Battista Alberti in the late 15th century. This had the 26 letters of the alphabet arranged in random order around the edge of a disk which, in turn, was pinned to the center of another, larger disk. The latter also had the 26 letters of the alphabet around its rim, and the two disks were free to rotate relative to one another.

To encipher a message, the disks would be rotated so that a particular letter on the inner disk aligned with a different letter on the outer disk. This combination would be revealed in the first letters of the message, say "Bj", to reveal that "B" on the outer disk should be set opposite "j" on the inner disk. The writer would then translate the message into the ciphered text by referring from one disk to the other. If he wanted to write the word "the" for example, he would look for the letter "t" on one of the disks and write down the letter with which it was aligned on the other disk, say "m." The second letter, "h," might line up with the letter "q" and the letter "e" with "f." In this way, the word "the" in the message would be written as "mqf," and so on. When the message arrived, the recipient would set his identical cipher wheel accordingly and carry out the entire operation in reverse.

This was clever enough, but the system still contained the weakness that whenever a word appeared more than once in the message, it would have the same transcription, so in our example "the" would always be represented by "mqf." But Alberti's masterstroke was to include a capital letter at intervals in each message, which was a signal to the recipient that the positions of the disks had to be changed to set the original calibration letter of the inner disk against the new letter on the outer disk, altering all the relationships between the plain text letters and those of the cipher text. After this change, for example, "the" would no longer be "mqf," but could be "sxz." The relationship could be changed twice or three times more, each time defeating the efforts of decryptors who might be looking for the most common letters or combinations.

Cipher squares and tables

Later, cryptologists used complex multi-alphabets to avoid even the small degree of repetition of combinations produced by Alberti's wheel. They switched to a new alphabetical relationship between cipher and plain text with every letter. Provided the recipient had an identical table with the alphabets set out upon it, deciphering the message was laborious, but straightforward, while deciphering it without the key was made much more difficult.

Other ciphers were invented that used word squares made up of vertical columns and horizontal lines of letters: the individual letter in the plain text was represented by the letter at the head of the column in which it appeared, and the letter at the beginning of the line in which it appeared. In other words, each letter in the original text would be represented by two letters in the ciphered text. Other schemes assigned different numbers to different letters, then carried out various mathematical operations on them before translating the result back into letters to produce the ciphered text. Anyone knowing the system would be able to carry out the whole operation in reverse to reveal the original.

Radio interception of messages

The value of ciphers expanded enormously with the invention of radio. Here was a wonderful new way of sending secret messages right around the world, with no danger of them falling into enemy hands through the capture of an agent. There was

65_ Télégraphe Morse.

ABOVE This 1910 picture (above) shows a message being received by Morse telegraph, decoded and written down for passing to the intended recipient.

BOTTOM Model of the sliding alphabet used to encipher and decipher the code known as Vignère.

only one snag, and that was that anyone else with a radio receiver could intercept the messages as well. Hence the need for a really secure cipher that would defeat an enemy's most determined efforts to break it.

Even here, though, there were weaknesses that clever eavesdroppers could exploit. Because of the huge amount of traffic to and from all manner of different places, which had to share a particular radio frequency, messages needed call signs to identify the sender and intended recipient. These had to be clear enough to be

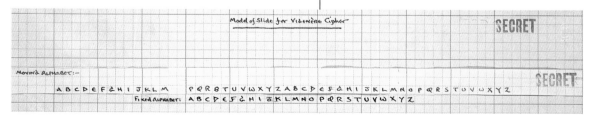

recognized without the need for someone to try to decrypt all the signals on a given frequency, and they often provided the first step in breaking a series of messages. Standard abbreviations used in all messages to indicate reception conditions and mistakes in sending a Morse message, and to ask for garbled words to be repeated added to the weaknesses of the most clever encryption scheme.

Another clue given away by any radio transmission was the location of the transmitter itself. The invention of the direction finder allowed a bearing to be taken to the transmitter by turning a directional antenna until the signal being listened to reached its maximum strength. By using two or more direction-finder stations in different places and plotting the resulting bearings on a map, it was possible to locate the transmitter, which would be at the intersection of the bearings. This could provide a very valuable clue as to the identity of the organization or the individual sending the messages.

Other methods of reading messages protected by the cleverest and most involved ciphers included capturing enemy ciphers and codebooks, either from individual agents or from enemy units overrun in combat. In many

ABOVE RIGHT Wartime propaganda poster promoting the BBC radio services to its French language listeners.

BELOW A typical 1940s radio receiver with its bulky, wood-veneered cabinet and large tuning dial.

cases, where agents were working in hostile territory, the use of direction finders enabled the counterintelligence organizations to capture them with ease. During World War II, the German direction finders operated by the Abwehr and the SD security services were able to work very quickly to track down radio operators, particularly when messages were complex and took time to transmit in full.

In crowded urban areas, a favorite trick was to cut off the electric power in different zones and note when a particular transmission was interrupted. In the countryside or the suburbs, it was easier to find the site of a transmitter, which had been revealed by cross-bearings, by looking for the antenna. However, radios were made smaller and more compact, allowing them to be carried to places where the approach of counterintelligence squads could be spotted in advance, while powering them by batteries made them less vulnerable to interruptions in electricity supply.

Another powerful defense against capture was to reduce the time spent in transmitting messages. With the

development of more sophisticated recording techniques, it was possible for an agent to record a long message at a very slow speed on a tape, then send it over a voice transmitter as a brief burst of noise. The message would be recorded at the receiving end, then played back at a lower speed to reveal the original message.

Changing codes and ciphers

The difficulty with all codes and cipher systems is the huge volume of communications traffic generated by a modern society, and the large number of recipients for messages used for government and military purposes, all of whom have to have copies of the right information to

ABOVE Members of the team broadcasting from Britain in French and Flemish to listeners in Occupied Belgium, a service which provided vital help for the Belgian Resistance organization.

decipher incoming messages and encipher outgoing ones. This means that if an organization only suspects its ciphers might be read, even partially, the correct solution is to replace the system altogether.

However, carrying this out when codebooks or cipher tables have to be delivered to units and individuals around the world may take months, or even longer in wartime. Changing the arrangements for security reasons may be an action that can only be taken on a long-term basis, and while the new systems are being distributed, the service using them will be handicapped severely. On occasions in the past, a partial distribution of a new cipher system has meant that some all-stations messages have had to be sent in both ciphers which, if picked up, would have given enemy code breakers an open door into the new system from the start.

Perhaps the ultimate unbreakable system, particularly when messages are relayed by voice rather than Morse code, is the employment of operators who speak a language of their own, not accessible or even identifiable to an enemy. During World War II, American units spread across the vastness of the Pacific kept in touch by using Navajo Indians as radio operators. They talked to one another in their own very complex language, in which a single word could convey the meaning of a whole sentence.

The Navajo language was virtually impossible for anyone to learn, other than by speaking it from birth, and only 28 non-Navajos were known to have succeeded in the task. Almost all of them were missionaries and anthropologists, and none of them had any

BELOW Some of the complex workings contained in a British intelligence manual used in helping to crack German ciphers.

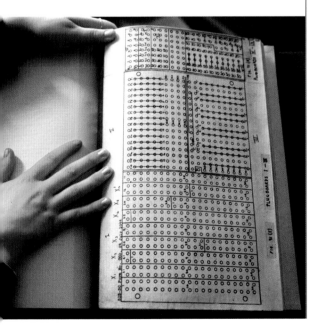

contact with Germany or Japan. Not only did even the most expert speakers reveal themselves as non-Navajo to the Navajos themselves, by the sounds they made when speaking the language, but also since the radio operators came from a tribe of some 50,000 people, many knew one another already. A combination of Navajo words, American slang and military code words produced messages that were completely secure from being understood by any Japanese radio operator who happened to intercept them.

More recently, the same principle was used by British Army units serving with the UN forces in Bosnia. Regiments like the Royal Welsh Fusiliers, which recruit a relatively high proportion of soldiers able to speak Welsh, have assigned them as radio operators to provide communications between different elements of a battalion on active service. Although Welsh is more widely spoken than Navajo, and easier to learn, the fact that there were unlikely to be fluent Welsh speakers among the local Serbs, Albanians and Croats helped keep tactical communications secure from interception.

Scrambling and unscrambling speech messages

Finally, speech communication can be scrambled. This involves passing the signal through circuits that alter the frequency of different tones in the speaker's voice. In some cases, the low frequencies are changed to high frequencies, and vice versa, to produce a high-pitched noise that is quite unintelligible as speech. Other scrambler systems increase all frequencies, but those at the top of the range are brought back into the overall signal as low frequencies. Some split the original signal into different frequency bands, then change them before reassembling the signal.

In the more sophisticated systems, the scrambling procedure can be changed at intervals of fractions of a second.

Time-division scramblers record a message using a set of recording heads that allow successive segments of the message to be shuffled in time order. A variation on this theme employs a movable head that travels back and forth over the tape on which the original message was recorded, so that the frequencies fluctuate rapidly up and down. The wartime telephone conversations between President Roosevelt and Prime Minister Churchill were also switched between different channels, so that anyone listening on one particular channel would find that the signal died away at regular intervals.

Nevertheless, German engineers were able to unscramble the signals in September 1941, after months of trial-and-error experimentation. Among the conversations revealed, they heard both men discussing the Italian withdrawal from the war in 1943, and in early 1944 remarks about the level of military activity in Britain which, to German ears, suggested that the cross-Channel invasion could not be long delayed. Shortly after that, the system was replaced by an even more complex version, which the Germans were unable to crack.

More recently, systems have been developed that add noise to the message so that when broadcast, it sounds like atmospheric interference, not worth even recording. When received by the intended station, however, the covering noise can be stripped away to leave only the message. Currently, research is being done into ways of using lasers to scramble and unscramble voice signals sent along fiber-optic communications lines, which might be intercepted by spies or counterintelligence agents en route.

Codes and ciphers depend on logical systems, understood by both the sender and the receiver. For example, one of the simplest of ciphers involves transposing the letters of the plain text. In a transposition cipher, the word "MIDWAY" might have the first pair of letters shuffled to positions 5 and 6, the second pair to positions 1 and 2, and the final pair to positions 3 and 4, producing the group "DWAYMI." Although the word has been changed, any experienced solver of crossword puzzles would have little difficulty in rearranging the anagram to find the true meaning. If the rest of the message was enciphered in the same way, it could quickly be broken down to reveal the correct text.

Consequently, most ciphers depend on substitution, or sometimes a combination of substitution and transposition. In substitution ciphers, the normal letters of the alphabet are replaced by cipher letters. In other words, the letter "a" in the plain text might become the letter "q" in the cipher text, "b" in plain text might become "v" in the cipher text, "e" might become "r," and so on. This would give no obvious clue as to the meaning of the sequences of the letters in the ciphered text, but the normal statistical rules would apply. Someone trying to break the cipher would look for the letter that appeared most frequently in a long

ABOVE A circular code which has been expanded into a table to make enciphering and deciphering easier and more straightforward.

LEFT Miniature one-time cipher pads found on two agents accused of spying for the Castro regime in London.

message. If the message was known to be written in English, and "r" was the letter that appeared most often in the cipher text, then it would be likely that "r" was the cipher equivalent of "e," and in this way, the plain text message could be deciphered progressively.

To make ciphers more difficult to break, a whole series of different cipher alphabets can be used, switching between them according to preset rules—changing for every letter or every 10th letter for example. This breaks up the statistical relationship

between letters, since an "e" in the plain text could be represented by an "r" on one occasion, a "v" on another, then a "b," and so on.

The system used in a particular cipher is the key to that cipher. In many ciphers, agents may rely on a keyword. British Special Operations Executive agents in Occupied Europe were given poems to use as keys for their messages. Unfortunately, if the Germans could deduce the poem used by an agent, they could decipher all of that agent's messages. As a stopgap, SOE agents had poems written specially for them, but they could be forced to reveal the poem if captured, and then the Germans could read all of that agent's previous messages.

Some espionage networks used book ciphers, where a letter would be identified in a message by its position on a certain page of a particular book, by line, word and individual letter number. This produced a cumbersome, but fairly secure, cipher, unless of course the book was found when an agent was captured. Later, SOE agents used "silks," cipher alphabets printed on silk, which would not crackle when hidden within the layers of a garment. Each strip of silk contained a key for the cipher alphabets to be used for a particular message, and when the message had been sent, the strip was torn off and destroyed. Because a different cipher was used for each message, the system was extremely secure, and agents could not be forced to reveal the keys, since they were random, unmemorable and different for each message. This was also the principle of the "one time pads" used extensively by Russian agents, where a different page of a preprinted notebook was used for each successive message.

Finally, codes are similar in basic principle to ciphers, except that whole words are coded rather than individual letters. This means that the sender and the receiver of a coded message must have a complete code dictionary, so that each word of a message can be looked up and translated into a group of letters or figures. Even so, occasionally patterns are revealed, which can be used to discern part of a message, since often words that are closely related can have very similar number or letter combinations. In other cases, ciphered messages may contain code groups in substitution for names of people or places, which otherwise may offer eavesdroppers a way into the message.

The Zimmermann telegram

After three years of bloody stalemate, the only hope the Germans had of defeating Britain in World War I was to use submarines to sink the merchant ships that carried the country's essential food and raw materials. Unfortunately, the rules of war required that a U-boat had to surface to stop and examine a merchant ship's papers before allowing the crew to take to the lifeboats, and only then could it sink the ship. This long and laborious process left the submarine vulnerable to Allied attack, so the Germans decided that the only way to achieve their objective was to introduce unrestricted submarine warfare. This meant torpedoing Allied and neutral merchant ships on sight from below the surface, but the policy risked goading America into entering the war on the Allied side, which undoubtedly would tip the balance the other way.

On January 17, 1917, the Royal Navy's intelligence headquarters, in Room 40 of the Admiralty Building in London, intercepted a German radio message that had been encrypted using a non-naval code. It contained more than a thousand sets of numbers, some of three figures, some of four and some of five. The only clue to its content was the opening group, 13042, which suggested a variation of the German diplomatic code, 13040. Fortunately, the Admiralty experts had a captured copy of the 13040 code, and a notebook containing details of all the variations they had been able to find on that basic code.

First they looked for the signature. The third group from the end was 97556, and usually high numbers like this were reserved for individual names. Checking through other messages, they found that it stood for Zimmermann, the Kaiser's Foreign Secretary. At the beginning of the message, they deciphered the words "Most Secret—for your Excellency's Personal Information." They knew that the message had been intended for Washington, so the recipient had to be the German ambassador to the U.S.A. Then they came across a code group for "Mexico" and another for "Japan."

The mystery appeared to be deepening. Slowly they uncovered text informing the ambassador that unrestricted submarine warfare would begin in two weeks, that this was expected to bring England to peace within a few months, and that

SECRET.

L.W.   February 24th 1 p.m. 1917.

HW 3/179

Balfour has handed me the translation of a cipher message from Zimmermann, the German Secretary of State for Foreign Affairs, to the German Minister in Mexico, which was sent via Washington and relayed by Bernstorff on January 19th.

You can probably obtain a copy of the text relayed by Bernstorff from the cable office in Washington. The first group is the number of the telegram, 130, and the second is 13042, indicating the number of the code used. The last but two is 97556, which is Zimmermann's signature.

I shall send you by mail a copy of the cipher text and of the decode into German, and meanwhile I give you the English translation as follows:-

"We intend to begin on the 1st of February unrestricted submarine warfare. We shall endeavour in spite of this to keep the United States neutral. In the event of this not succeeding we make Mexico a proposal of alliance on the following basis:-

'Make war together - Make peace together'. Generous financial support and an understanding on our part that Mexico is to reconquer the lost territory in Texas, New Mexico and Arizona. The settlement in detail is left to you.

You will inform the President of the above most secretly as soon as the outbreak of war with the United States is certain and add the suggestion that he should on his own initiative invite Japan to immediate adherence and at the same time mediate between Japan and ourselves.

Please call the President's attention to the fact that the ruthless employment of our submarines now offers the prospect

of

the German government—in the event of America becoming hostile—would propose an alliance between Germany and Mexico. But 30 of the number groups resisted all their efforts for weeks on end.

When they finally broke the core of this message, they were astonished. It revealed that if Mexico would ally herself with Germany, the Germans would help her "regain by conquest her lost territory in Texas, Arizona and New Mexico," and would try to negotiate a treaty with Japan, then part of the Allied cause, but a potential threat to U.S. interests.

The information was sent to the United States, and Congress immediately passed a bill for the arming of American merchant ships. However, some members wondered if the text might be an Allied plant to persuade America to enter the war. The Americans had also intercepted the coded text, however, and it was decoded for them by the British experts to verify the information they had been given. To prevent the Germans from learning that their code had been read, stories were planted in the press by the Admiralty that American agents had found the information on a German agent attempting to cross the border into Mexico, and the British newspapers enjoyed criticizing the ineptness of the British counterintelligence services compared with the achievements of the Americans.

The Germans concluded that the leak had indeed occurred in Mexico and decided not to change the codes. But the final stamp of authenticity was provided by Zimmermann himself, who made a public admission that the telegram had been sent on his authority. The news galvanized American opinion, particularly in the inland states most affected by the German plans, which previously had seen the war as a purely European affair. A seismic shift in public attitudes persuaded President Wilson to ask Congress for a declaration of war on April 2. By the careless use of a compromised code for a message of such awesome sensitivity, Germany had doomed herself to defeat.

Le Petit Journal

ADMINISTRATION
(1, rue LAFAYETTE, 61)
10 CENT.   SUPPLEMENT ILLUSTRE   10 CENT.   ABONNEMENTS
28me Année   Numéro 1.366
DIMANCHE 25 FÉVRIER 1917

M. WOODROW WILSON

PRÉSIDENT DE LA RÉPUBLIQUE DES ÉTATS-UNIS

ABOVE US President Woodrow Wilson featured on the cover of a French magazine in February 1917, just weeks before he requested a declaration of war from Congress to bring America into World War I.

OPPOSITE BOTTOM Text of a United States diplomatic telegram sent from London to Washington relating to the Zimmermann message and the subsequent entry of America into the war.

OPPOSITE BACKGROUND Woodrow Wilson in 1921.

Rommel's American agent

ABOVE Field-Marshal Erwin Rommel, German commander in chief in North Africa was dubbed the "Desert Fox" for his cunning, but he based his astute decisions on intercepted information from an American officer's reports to Washington.

During the fighting in the North African desert, between 1941 and 1943, the German General Erwin Rommel outmaneuvered his British opponents so often that he became known as the "Desert Fox." Often he seemed to know just what the British were going to do, allowing him to time his counterstroke for maximum effect, time and again.

In fact, his information was coming from a highly placed American officer, Colonel Bonner Frank Fellers, who had been posted to Cairo as a liaison officer with the British forces. He studied the desert battles and sent coded, very detailed reports back to Washington on British tactics, morale, the arrival of reinforcements and the plans for future operations. Unfortunately, the cipher he used was one that the Germans had broken, and Rommel's staff were intercepting every signal, so the general had the clear text within hours of it having been sent.

Fellers' messages provided priceless intelligence. They revealed specific details, like planned commando raids on German desert airfields, and descriptions of British defenses and fortifications. The effect of this information was to change the balance between the two armies. Rommel could outflank heavily fortified areas and attack them from the rear. The commando raids were defeated, and the aircraft they failed to destroy were used to attack supply convoys to Malta, sinking so many ships that the island remained under siege for another five months.

In the end, this fatal leak was revealed when Hauptmann Alfred Seebohm, who commanded Rommel's radio interception unit, was killed in action and his records fell into British hands. At the same time, a German prisoner told his captors that the Germans were reading Fellers' code. The British notified the Americans, Fellers was recalled and the codes were changed. Rommel lost his window on Allied plans and intentions at the very moment that "Ultra" handed the advantage to the British. With the loss of this vital information, Rommel's predicted victory was transformed into final defeat.

BELOW After Rommel—seen directing an attack against the British—lost his priceless source of information, his succession of victories came to an end.

Jefferson's Cipher Wheel

The American statesman Thomas Jefferson developed an ingenious mechanical ciphering and deciphering system during the 1790s, which he referred to as a wheel cipher. It comprised a central shaft that carried a set of rotating rings, each scribed with the 26 letters of the alphabet. To send a message, each of the rings was rotated so that the letters of successive rings spelled out the message along the length of the shaft. Looking at any of the other lines on the shaft produced a completely indecipherable string of letters. Any of those lines of letters could be used as the cipher text of the message to be sent, since no one could read the plain text from them.

The recipient arranged the rings on an identical shaft so that the letters reproduced the scrambled message. Simply looking around the shaft would reveal the plain text at some point, since only one of the lines of letters would be readable. Variations on the basic principle employed greater numbers of rings, allowing longer messages to be sent and deciphered, although a really long message could be split into selfcontained sections. Furthermore, by varying the order in which the alphabet was scribed on each of the rings, and changing the order in which the rings were fitted to the shaft, an astronomical number of potential combinations could be created.

The Jefferson device was simple and effective, and very secure for its time. However, the details were only found among his papers in 1922, so it appeared that he did nothing to publicize it or recommend its use. Nevertheless, in that same year, the US Army adopted an independently developed ciphering device that was based on the same basic principle as Jefferson's wheel. It was also used by government agencies and proved remarkably resistant to code breaking. The US Navy continued to employ it until well into the 1970s.

Finding the convoys

ABOVE German field service badge worn as a decoration by troops who had taken part in the Narvik campaign of 1940.

During the early years of World War II, the German radio interception and cipher breaking service, the B-Dienst, developed an astonishing expertise at reading Royal Navy codes. When the British decided to lay mines along the Norwegian coast to prevent vital iron ore shipments from neutral Norway reaching Germany, Hitler made a plan to invade that country in April 1940, realizing only too well that the troopships carrying the invading force, intent on landing in southern Norway, would be open to naval attack.

However, the B-Dienst intercepted signals that revealed the British intention to occupy the strategic port of Narvik in the far north of Norway. The Germans sent out a force of warships, which headed north as if to threaten the British landings, and the British reacted, as the Germans hoped they would, by sending all their naval forces northward. The decoy force vanished into the mists and snow squalls, and in the south the troop transports landed the occupying army without any Royal Navy opposition at all.

But the B-Dienst's greatest achievement was the breaking of the BAMS—Broadcasting for Allied Merchant Ships—code. The German raiding ship *Atlantis* was a heavily armed, high-speed freighter that was preying on lone British merchant ships in the Indian Ocean during the summer of 1940. On July 10, she encountered the *City of Baghdad* and opened fire. The latter's crew took to the boats, and the Germans boarded their prize, to find a set of BAMS codebooks and tables. Although these tables were changed regularly and the seized copies were out of date, the radio operator of the *Atlantis* was able to deduce enough of the changes to be able to read current messages. This enabled the raider to sail straight to other isolated targets in the broad wastes of the southern oceans.

More captures led to further inroads into the British cipher system, and all were relayed back to Berlin, and from there onto the U-boat command in Western France. By 1941, the B-Dienst were reading messages to Allied convoys from the Commander in Chief Western Approaches, based at Liverpool, giving the ships their positions, timings and routes through the areas where the U-boats operated. They were even able to read British signals alerting their forces to the positions of all known U-boats at sea, so they could eavesdrop on what their enemy knew of their whereabouts!

Later, though, the tables were turned with a vengeance. When the German naval Enigma ciphering system was broken by the Allies, they knew exactly where the U-boats were, from their own signals reporting back to base. At one time Allied convoys were being rerouted to avoid the locations of the U-boats, while the U-boats were being diverted in turn to match the new courses and speeds of the convoys in a macabre dance of death, made possible by both sides reading one another's signals. But in May 1943, when the advantage was shifting at last in the Allies' favor, the British ciphers were changed, and the B-Dienst had to begin the laborious process of breaking the new codes all over again. By the time they began to succeed, the U-boats had already been defeated.

Hoodwinking the Japanese

During the months leading up to the attack on Pearl Harbor, American cryptanalysts had succeeded in breaking a number of Japanese ciphers. A few hours before the attack on the Hawaiian naval base, one of these revealed that Japanese diplomats in Washington were being instructed to burn their codebooks and destroy most of their cipher machines, a clear prelude to the outbreak of war. Unfortunately, none of the deciphered messages contained any real clue as to where the Japanese intended to strike. Pearl Harbor was alerted, along with all other American bases in the Pacific, but delays in communications meant that the vital warning arrived just after the Japanese struck.

In many ways, the Japanese surprise attack was, from their point of view, brilliantly successful, but in one crucial aspect it was a failure. While their aircraft crippled the American battleships, they missed the all-important American aircraft carriers, which were safely out at sea. By the following spring, the Japanese had conquered Malaya, the Dutch East Indies (modern Indonesia), the Philippines and parts of New Guinea and Burma. Only the American carriers were able to mount a challenge to this overwhelming wave of victories.

Their first success came when US Navy cryptographers deciphered a message in the JN25 naval cipher in April 1942, indicating that the Japanese planned to land troops in the Solomon Islands and at Port Moresby, on the southern coast of New Guinea. Both operations would take the Japanese fleet across the Coral Sea, but when the opposing fleets searched for one another, it took three days before they finally made contact. The Japanese and the Americans each lost a carrier, but the former failed to capture Port Moresby, and their invasion fleet withdrew.

The crucial question was where would the Japanese strike next? They could attack west toward India, south toward Australia or east toward Hawaii, and it was essential that the

ABOVE Following the victory of the Battle of Midway and the loss of the carrier *Yorktown*, the Americans built a new carrier bearing the honoured name which was launched in 1943.

OPPOSITE ABOVE The crew of the *USS Yorktown* on parade.

OPPOSITE BELOW Japanese destroyers in line ahead formation as the advance guard of the fleet.

Allies knew which. Once again, American cryptographers concentrated their efforts on messages in JN25 and, sure enough, they were able to decipher partial messages that hinted at another, even larger operation. Commander Joseph Rochefort, head of the Navy Combat Intelligence Office on Hawaii, was convinced that the target was the American base on Midway Island, located on a direct line between Japan and Hawaii, but there was no definite evidence. All the Americans knew was that there were two objectives, referred to simply by code letters. "AL," they were convinced, stood for the Aleutian Islands in the North Pacific as a diversion, but the main objective was "AF." Did this stand for Midway, for Hawaii or for the Pacific coast of America itself?

Then Rochefort had a brilliant idea. The Midway base depended on a filtration plant for its drinking water, and he gained official permission for Midway to send a signal in plain language to Pearl Harbor, reporting problems with the plant and asking for water to be delivered by tanker. The radio interceptors kept watch on all Japanese frequencies and, on May 12, their vigilance was rewarded by a signal to Japanese warships which, when deciphered, read "AF is short of water."

The Japanese main objective was clearly Midway, and this enabled the US Navy to locate its carrier force where the oncoming enemy would least expect it. Convinced that, as with Pearl Harbor, they had achieved complete surprise, the Japanese approached their target on June 4, 1942. Although both sides were dogged by tactical errors and a measure of bad luck, in just five minutes, between 10:20 and 10:25 am, the balance of naval power in the Pacific was transformed forever. For the loss of the carrier *Yorktown*, the Americans sank no fewer than four of Japan's largest and most powerful carriers, a victory that would not have been possible without the skill of the code breakers.

Beating the Machines

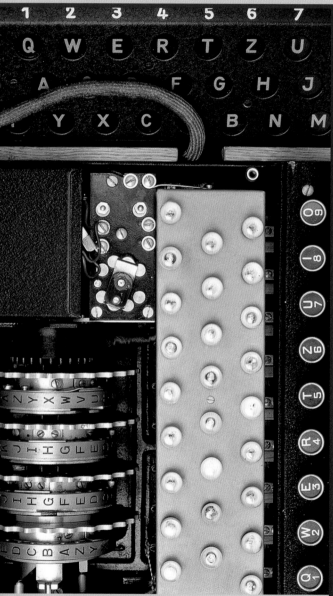

Compared to lone agents operating under hostile conditions, major institutions, such as a country's diplomatic corps and armed forces, have different priorities when it comes to preventing their messages from being read by an enemy.

To begin with, the huge volume of traffic they have to send to keep in touch on a regular basis is routinely intercepted and recorded by counterintelligence agencies around the world, keen to decipher every scrap of information possible. On the other hand, carrying out regular changes to make transmissions immune to cipher cracking becomes a very complex task when every embassy, every consulate, every warship, every airfield and

BELOW These miniature code books of one time pads offer even greater security than the most complex cipher machines but take time to distribute and may be captured by enemy agents.

ABOVE This World War II German Enigma cipher machine has four rotors, of the type issued to the German Navy, making its messages more difficult to break than those sent by Army and Luftwaffe three-rotor machines.

every army unit has to be given copies of the new ciphers in time for the changeover.

Between the two world wars, inventors in Sweden, Holland, Germany and the USA attempted to solve the problem with machines rather than paper cipher tables. In Germany, an engineer and inventor named Arthur Scherbius filed the first of a series of patents for a cipher machine in the spring of 1918. This was based on a normal typewriter, which had been modified so that striking a key turned a disk-shaped rotor carrying 26 electrical contacts, one for each letter of the alphabet. As the person typing out the plain text message struck successive keys, the rotor would continue to turn, bringing different contacts in touch with the machine's internal wiring.

Known as the Enigma machine, on later versions each character of the cipher text was displayed, not on paper, but by the illumination of one of 26 lamps on the top of the machine, each indicating a letter of the alphabet. The position of the rotor determined which lamp lit up, and since this changed as each key was pressed, the relationship between the plain text and the cipher text altered with each character typed. If the operator typed the same letter several times, for example, the cipher text would show a different character each time. It was equivalent to using a different cipher alphabet for each letter: very effective, but an enormously laborious task if done by hand.

However, this was merely the basic principle. With a single rotor, the relationship between the plain text character and the cipher text character would be reproduced every 26 key strokes. But Scherbius also proposed a set of rotors that would work in series, like the changing figures on a vehicle's mileage recorder. When the first rotor had completed a full revolution, gearing would advance a second by one division, again out of 26. Using a third rotor in the same way increased the number of characters in a

message before the original relationship was repeated to a total of 26 x 26 x 26, or 17,576. Later, a fourth rotor was added, although this did not actually turn, and it had 13 contacts instead of the normal 26. This was used as a reflector, sending the electrical signal generated by pressing a key, which had reached it through the different settings of the other three rotors, back through them, but on an entirely different path to complicate matters still further.

This meant that for the recipient of a message enciphered in this manner, it was essential to know the order in which the removable rotors were fitted into the machine (since they had different internal wiring, they were not simply interchangeable, which would have been a potential weakness) and also the starting positions of the rotors before the first key was pressed. Provided with this information, all the receiving operator had to do was set up the machine and type in the

BELOW Two of the rotors used on an Enigma machine showing the electrical contacts between adjacent rotors which added to the complexity of the cipher. These rotors became prize catches for any interceptors of German Navy craft.

ciphered text as received by radio. Noting the letters revealed by each indicator lamp would produce the original plain text.

After World War I, the Germans had realized that, having captured vital codebooks, the British had been reading their naval and diplomatic codes, and had put the information to good use, as had been the case with the Zimmermann telegram (see Chapter 4). The attraction of the Enigma machine was that it did not rely on codebooks, and even if a machine was seized, it would be useless without details of the rotor settings.

Furthermore, the versions adopted by the German Navy and Army were provided with a selection of rotors, all with different wiring, so it was necessary to choose the correct three rotors as well as know the settings before it was possible to decipher a message. Later, the Army specified the addition of a plugboard containing 26 sockets, which could be connected in pairs using short cables. In

practice, six of these cables were used to interconnect 12 letters, but it added another layer of complexity: the connections were also needed to decipher Enigma messages.

The result was a machine that its users were convinced would make their signals totally unreadable by an enemy. Having learned a bitter lesson in World War I, the Germans were sure that the Enigma system had closed that deadly loophole once and for all. Indeed, so certain were they of its infallibility that they could not conceive of the system being breached. When the Allies showed worrying signs of knowing too much about developments and movements that had been detailed in Enigma signals, the Germans adamantly refused to believe that the system was at fault.

The supreme irony of the situation was that their attempt to ensure that the British code-breaking successes of World War I could not be repeated, handed their enemies a much

we here highly resolve that these dead shall not have died in vain . . .

REMEMBER DEC. 7th!

more valuable, and ultimately much more damaging, victory, which would not have been theirs had more traditional ways of enciphering and deciphering messages been used from the outset.

The Germans offered the Enigma system to their Japanese allies, who used the same principle for important diplomatic messages, together with the characters of the Western alphabet. However, their Alphabetical Typewriter actually printed out the cipher text using a second machine connected to the plugboard of the first. Like the Germans, the Japanese were impressed with the

TOP US Government poster issued in 1942 to raise morale and promote recruitment after the Japanese attack on Pearl Harbor.

OPPOSITE The smoke from burning oil tanks contrasts with the violent explosion as the magazine of the destroyer *USS Shaw* explodes during the Pearl Harbor attack.

ABOVE General George C Marshall, Chief of Staff of the US Army, visits the Normandy fighting after the D-Day landings in June 1944.

mathematics of the machine, and also assumed that it was invulnerable to any American cipher crackers who might record their radio signals. But just as the British were able to break the Enigma messages to produce a flood of vital intelligence information for most of the war, their American opposite numbers were to achieve the same success against the Japanese, with radical effects on the course of the Pacific conflict, although one of the greatest catastrophes was that they didn't manage to decipher the message that would have given them advance warning of the attack on Pearl Harbor in 1941.

ABOVE Damaged aircraft at Pearl Harbor. The attack was so swift and unexpected that most of the US defenses were destroyed before even leaving the ground.

OPPOSITE TOP LEFT The feeling of shock and fury felt by the American people after Pearl Harbor was swiftly followed by a sense of retribution, encapsulated in this US Government poster.

OPPOSITE TOP RIGHT Military personnel looked on in horror as the first wave of Japanese attacks began.

OPPOSITE The destroyer *Shaw* is ripped apart by an immense explosion when her magazine is ignited by enemy fire during the Japanese attack.

Breaking Enigma—and keeping the secrets

How was Enigma broken? Apart from the advances provided by seizing German tables showing the daily settings of naval machines, much of the breaking of the ciphers was done by patiently chipping away at the protective system. Despite the awesome number of combinations made possible by the different rotors and the plugboard, there were certain weaknesses in the system that actually reduced the number of possibilities. For example, any letter in the plain message text could be represented by any other letter of the alphabet, with the exception of itself. Secondly, there was a reciprocal relationship between the plain text letter and its cipher equivalent—if the plain text letter "s" became a "Q" in the cipher text at a particular rotor position and order, and with a given set of plugboard connections, then the plain text letter "q" would be represented by "S" in the cipher text at the same settings.

In addition to these relationships, it was possible to try certain combinations of letters for messages known to originate from a particular source, identified by radio direction finding. Army and Luftwaffe Enigma messages always began with six code letters, which gave the recipient additional information about the settings used,

but after that usually would be addressed to a particular unit or individual. When the letters of a possible address were written above the letters of the ciphered message, the Bletchley Park code breakers would look for letters that showed a particular loop; for example, a plain text "c" over a cipher text "Q," with later a plain text "q" over a cipher text "B," and then a plain text "b" over a cipher text "C," giving a loop of "CQBC."

If the diagram showed one of these loops, it was used in setting the "bombe" computer, which then would work through each of the possible settings of the machine. If a given setting was incorrect, the wiring of the "bombe" would be activated and light up most, if not all, of the 26 indicator lamps. If there was a possibility that it was correct, only one lamp would be lit, and this relationship could be applied to the message to see if it produced any recognizable German text. Basically, the process was one of trial and error, carried out by fast, reliable machines that could reduce months of painstaking work to as many minutes.

Other loopholes included "kisses," Bletchley Park slang for two versions of a signal being sent at the same time, one enciphered by Enigma and the other by a lower grade

ABOVE AND BACKGROUND
The Colossus electrical digital computer was developed in 1943 to help with cracking the German Enigma signals by running through countless possible combinations to reveal the plain text.

LEFT The codebreaking rooms at Bletchley Park.

LEFT The four-rotor German Enigma machine was also issued with boxes of alternative rotors having different connections to provide an additional level of security.

cipher that they could read with relative ease. For example, the low-grade Dockyard Cipher was regularly read by the British and occasionally would offer a way into naval Enigma. It was also used to reveal details of the German Navy's grid for identifying positions in the Atlantic. RAF aircraft were instructed to lay mines in a particular area, then Bletchley Park would be able to read German signals that gave the grid references to areas now closed because of the mines.

This hard-won information needed special protection, and the Allies were always nervous of acting on intelligence revealed by Enigma messages when that same information could not have come from other sources. For example, at a relatively early stage in the war, they were able to read messages detailing the return of German surface raiders to their home ports in Occupied France. Sinking these raiders would have removed a threat to Allied shipping, but if all of them had been intercepted by warships in the huge tracts of the Atlantic, there was a risk of giving away the fact that German signals were being read. Finally, the decision was taken to attack all of them except two, to conceal the source of the information.

During the war in North Africa, Rommel's forces were supplied by sea from Italy. When Enigma intercepts revealed the sailing times and routes of the tankers, the British were concerned that sending torpedo bombers from Malta to sink them and ensure that Rommel's Afrika Korps ran out of fuel and ammunition would give away the fact that they were reading these messages. So, after

sinking the ships, they sent a message in a cipher they knew the enemy could read, congratulating a non-existent resistance group in Italy and thanking them for the information, which had enabled the supply ships to be intercepted and sunk.

Later, at a crucial stage in the Battle of the Atlantic, the Germans developed a means of keeping their submarines operating for much longer periods in distant, target-rich waters, like the US coast and the Caribbean. By employing "milch-cows"—supply submarines that carried extra fuel, provisions and torpedoes—they were able to increase the effectiveness of the operational U-boats in areas where they could do most damage.

The positions where the U-boats and their supply submarines were to meet were sent by Enigma ciphers, and when these were read by the Allies, the possibility of countering this new German tactic arose. After much heart searching, it was decided that the benefits outweighed the risk. At that stage of the war, there were several task groups at sea, each based around a small escort carrier. If the carrierborne aircraft searched the area where a "milch-cow" was known to be, spotted it, and attacked it, the encounter could appear to be the result of a chance sighting. Between June and October 1943, nine of the 10 German tanker submarines, each capable of supplying an entire wolf pack, had been sunk. A month later, the Germans withdrew from the convoy routes altogether, a defeat in which being able to read German signals had played a major part.

LEFT The rear view of Colossus.

TOP RIGHT The task of checking and cross-checking the messages and codes that cane in was an enormous task.

RIGHT The groundbreaking electronics of Colossus.

Cracking the Purple ciphers

ABOVE Code breaking equipment at its most basic: headphones to intercept radio messages and pads to note down the text for later analysis.

BELOW The wreckage of a B17 Flying Fortress on a Hawaiian airfield after the Pearl Harbor attack.

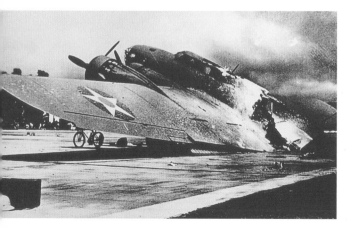

The Japanese first used a modified form of the Enigma machine for communicating with their embassies and legations overseas, and American code-breaking specialists referred to this as "Red." Later, the machines were replaced at major embassies by an upgraded version, which the Americans called "Purple." Because the machines were based on the European model, they used Roman characters, rather than Japanese script. Furthermore, they could not send numerals or punctuation marks, like commas, colons and periods, which had to be given three-letter codes instead.

In a brilliant cipher-cracking coup, American experts had intercepted prewar radio messages prepared by the Red machines and had broken the ciphers, employing a detailed knowledge of key phrases and modes of address used in formal diplomatic signals. Sometimes, the Japanese would send a message to their embassy in Washington containing the text of a note to be delivered to American diplomatic officials, and in this case, the Americans would have both the ciphered text and the plain text. This revealed the cipher system used, enabling them to read other messages prepared with the same settings.

Eventually, the Americans built their own analogs of the Red and later Purple machines, which employed a more complex cipher system. Often, the Japanese would have to send the same signal to their embassies and legations around the world, some via Red machines, and others through Purple machines. When the Americans deciphered the Red text, it gave them the messages that were also the subjects of the Purple transmissions, opening a route into deciphering those messages too.

Nevertheless, in the end, it took more than a year-and-a-half of dedicated, round-the-clock work before the US Army's Signal Intelligence Service deciphered the first full Purple message in August 1940. After that, they were able to read more and more of this vital diplomatic traffic, including a message that almost gave advance warning of the catastrophe at Pearl Harbor.

The crucial message was picked up by a US Navy radio-intercept station near Seattle, at 1:28 on the morning of December 7, 1941. When deciphered at naval headquarters in Washington, it revealed instructions for the Japanese ambassador to hand a final note to the US government no later than 1 o'clock Washington time that same day. It was then just after 5 o'clock in the morning, and the full text of the note comprised 14 parts, transmitted from Tokyo since noon the previous day. It announced the breaking off of diplomatic negotiations which, when added to earlier messages ordering the destruction of codebooks and the burning of secret documents, presented an ominous picture.

The complete message had been deciphered by 8 am Washington time. By 10:20, a copy had reached Lieutenant Commander Alwin D. Kramer, a Japanese language expert who realized what the 1 o'clock deadline implied: a surprise attack, often used by the Japanese immediately after a declaration of war, to gain maximum advantage over an enemy. One o'clock in Washington would be 7:30 am at the US Pacific Fleet base at Pearl Harbor. But there was nothing in the Japanese message to indicate where an attack might be delivered.

Meanwhile, the Japanese cipher clerks in the Washington embassy were also decoding the messages, while in the predawn darkness of the Pacific, the Japanese carriers were less than 300 miles from their target, maintaining strict radio silence. In Washington, DC, the deciphered message reached General Marshall at the War Department. He was reluctant to use his scrambler telephone link to Hawaii in case the Japanese could eavesdrop on the line, so he ordered the signal to be enciphered and radioed to Hawaii, the Pacific coast of the USA, the Philippines and the West Indies bases. Unfortunately, the radio station at Honolulu was experiencing severe atmospheric interference, so the only way to relay the message was through the powerful RCA broadcasting station in San Francisco, via the local Western Union cable office.

The signal was finally received by the RCA station in Honolulu at 7:33 local time on the morning of Sunday, December 7. At that moment, the first wave of the Japanese carriers' strike aircraft was showing up on U.S. Army radar, less than 40 miles from the island. Meanwhile, the message was placed in a tray to await hand delivery—the teleprinter link to the army base had been set up the day before, but was not operating, since it had not been tested.

Finally, at 7:55 am, the first Japanese bomb struck the seaplane ramp at the southern end of Ford Island, in the center of the Pearl Harbor naval base. The Japanese embassy in Washington was still struggling to decipher the complex series of messages, so the declaration of war, which should have preceded the attack, was never delivered. The same deciphered message, which should have alerted the American defenses, arrived after the attack had begun. But it was the apparent deliberate omission of a formal declaration of war that convinced America that Japan should be punished for what was seen as the ultimate act of treachery.

ABOVE The enduring image of Pearl Harbor as flames engulf the hulk of the battleship *USS Arizona*.

BACKGROUND Smaller ships too fell victim to the Japanese attack, like the destroyers *Cassin* and *Downes*.

Breaking the naval Enigma

BACKGROUND Bletchley Park, the country house in Buckinghamshire that was the base for the British teams working to decipher German signals and codes.

BELOW Imitation is the sincerest form of flattery—a British Type X Mark III cipher machine showing similarities to the Enigma.

Although the Allies knew exactly how the German Enigma machine worked, without the details of the initial settings of the rotors and the plugboard, they were powerless to read the messages. To speed the task of trying the vast numbers of possible cipher alphabets, the British code-breaking organization, based at the old country house of Bletchley Park in Buckinghamshire, developed a computer.

The team included university academics, adept chess players, crossword-puzzle setters and solvers, and mathematicians. One of their brightest stars, the eccentric, but brilliant, Alan Turing produced a machine called a "bombe." Eventually, relays of these machines, running at full speed to check thousands of different cipher alphabets every minute, would be used to crack the ciphers protecting the mass of German signals intercepted each day.

The German Navy gave its ciphers the greatest protection. German ships carried tables of Enigma settings, printed on water-soluble paper, which disintegrated when wet. Backup copies were held in a special protective envelope, which let in water if dropped into the sea when the crew was faced with capture. With the threat of U-boat attacks on Allied convoys being of prime concern, it seemed that the fastest way to read German Navy messages would be to capture the machine settings.

In February 1940, the German submarine U33 was attacked by a Royal Navy minesweeper, *HMS Gleaner*. The submarine surfaced and the crew were picked up by the British. Their wet clothes were taken away and they were given warm blankets,. Machinist Fritz Kumpf had put the Enigma rotors in the pockets of his pants, then forgotten about them. When he got his pants back, the pockets were empty. The British had the rotors.

The next breakthrough occurred on March 1, 1941, when commandos attacked the Lofoten Islands, off Occupied Norway. One of the escorting warships, the destroyer *HMS Somali*, spotted a small armed German trawler, the *Krebs*. The trawler opened fire, but was wrecked by three direct hits. When a naval party boarded the trawler to rescue survivors, they found an Enigma machine, a full set of rotors and the February settings table, enabling Bletchley Park to read German naval signals for the first time.

Two months later, a converted trawler called the *München* left the Norwegian port of Trondheim to patrol east of Iceland and radio vital meteorological data back to Germany. Her messages were read by Bletchley Park, and Harry Hinsley of the naval section suggested that the navy attempt

to capture her machine and settings tables. Three cruisers and four destroyers sailed from Scapa Flow, in the Orkneys, and made for the *München*'s position. One of the destroyers, the *Somali*, found the target and opened fire. The Germans threw their Enigma machine overboard and tried to escape, but were quickly caught.

A prize crew, from the cruiser *HMS Edinburgh* and including Captain Haines of naval intelligence, boarded the trawler and discovered the Enigma settings tables for May and June. The papers went to Bletchley Park, making it possible to read an increasing number of German messages. Then, two days later, the destroyer *HMS Bulldog*, escorting convoy OB318, depth-charged the submarine U110, forcing it to the surface. A boarding party seized a huge haul of documents and the boat's Enigma machine.

One problem remained. The Germans changed their machine settings every month, and soon it would be necessary to capture a new set. Another German weather ship, the *Lauenburg*, was heard on patrol northeast of Iceland in June 1941, and two cruisers and three destroyers left Scapa Flow on June 25 to intercept the vessel. The destroyer *HMS Tartar* sighted the *Lauenburg* at 6:59 on the evening of June 28, when the German trawler emerged from behind an iceberg. *Tartar* and her sister ship, *HMS Bedouin*, together with the cruiser *HMS Nigeria* opened fire, trying not to score a direct hit. The crew took to the boats, but two remained behind, throwing the Enigma machine overboard and burning secret documents in the ship's furnace.

Nevertheless, a British boarding party retrieved 13 mail sacks of documents, including the July setting tables, keeping the window on German secrets open for a little longer. But at the end of August, a joint operation between the Royal Air Force and the Royal Navy led to the capture of U570, which was sailed back to Britain. On board was an Enigma machine with spaces for four rotors, which would make the ciphers much more difficult to break. Then, at the beginning of 1942, all German naval messages became completely unreadable.

The blackout lasted until the autumn, when U559 was forced to the surface in the eastern Mediterranean by depth charges from the destroyer *HMS Petard*. Once again, a British boarding party entered the damaged submarine and retrieved Enigma papers, although the destroyer's first lieutenant, Antony Fasson, and Able Seaman Colin Grazier went down with the submarine when it sank suddenly. Each was awarded the George Cross posthumously, but their sacrifice ensured that by the end of the year, Bletchley Park was reading U-boat messages again.

ABOVE Alan Turing's original notes on the workings of the Enigma machine helped the team develop an understanding of its weak points which enabled the ciphers to be attacked and broken.

BELOW A four-rotor German Enigma machine used for the naval ciphers, with a remote lamp board allowing a second user to read the text produced by the machine.

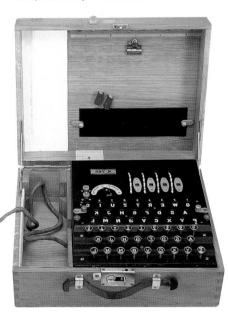

"Ultra" in the Pacific

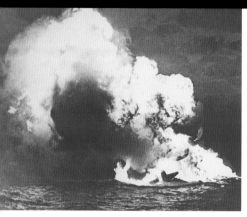

ABOVE The end of Japanese air power, as another aircraft crashes into the sea during the "Great Marianas Turkey Shoot" of June 1944.

BELOW The American aircraft carrier *USS Enterprise* damaged at the Battle of Santa Cruz in October 1942, a victory made possible by Ultra intelligence.

Following the unexpected Japanese attack on Pearl Harbor, which catapulted the United States into the war, and the battles of the Coral Sea and Midway, which threw the Japanese off balance, the Americans were able to capitalize on their brilliant coup in breaking the Japanese ciphers. First, though, they had to break back into the JN25 naval cipher, which had been altered following references in American newspapers to the deciphering of messages that revealed the enemy's intentions to attack Midway.

Until they succeeded, a combination of direction finding and traffic analysis gave vital information on the whereabouts and movements of Japanese naval forces, essential during the long fight for the island of Guadalcanal, in the Solomons, and its vital airstrip of Henderson Field. By late October 1942, they had been partly successful, and an "Ultra" (the cover name for the Allies' cipher-breaking efforts, meaning "ultra top secret") bulletin warned the fleet of another Japanese naval offensive directed against Guadalcanal. Although details were lacking, the American carriers *Hornet* and *Enterprise* headed north of the Santa Cruz Islands. Then a patrol aircraft sighted two Japanese carriers, 350 miles to the northwest. The ensuing Battle of Santa Cruz led to the sinking of the *Hornet*, while the *Enterprise* was damaged. In return, two Japanese carriers had been crippled, but far worse were the terrible losses in experienced aircrew that the Japanese could ill afford.

However, the greatest coup of the Pacific War was set in train by two corvettes of the Royal New Zealand Navy, *HMNZS Kiwi* and *HMNZS Moa*, which attacked the Japanese submarine I-1 off the western tip of Guadalcanal on the evening of January 29, 1943. When the submarine surfaced, *Kiwi* rammed it and forced it aground on a nearby reef. The crew escaped, and *Moa* sent a boarding party, which was able to seize the submarine's cipher tables. This success had been made possible in the first place by the deciphering of Japanese messages referring to submarines being sent to supply their beleaguered forces on Guadalcanal, which accounted for the two warships being in the right place when I-1 showed up.

Although the cipher books were out of date by the time they reached Pearl Harbor, they enabled American cryptographers to read earlier messages that previously had defeated them. Thus they were able to learn much more about Japanese units, organization and signals procedures. This kind of information was priceless in enabling them to break new Japanese messages, including one that was deciphered on April 14, 1943, revealing

ABOVE A Japanese cruiser is left burning and helpless at the Battle of Leyte Gulf in October 1944, when Japanese intentions and movements were revealed by reading their signals.

BACKGROUND Following the Leyte Gulf victory, American troops are able to land and liberate the Philippines.

that the Japanese naval commander-in-chief, Admiral Yamamoto, was due to undertake an airborne tour of bases in the Northern Solomons four days later.

One of the stops on Yamamoto's route was the island of Ballale, which was just within range of fighters from the airfield on Guadalcanal. The precision and the detail of the timetable in the Japanese signal enabled the Americans to send a force of 16 P38 Lightning long-range fighters to intercept the Japanese admiral and shoot him down. At 9:35 on the morning of April 18, exactly on schedule, they found Yamamoto's flight of two Japanese bombers, escorted by six fighters. The American pilots shot down both bombers (and a third that turned up in the middle of the action) together with three of the escorting fighters. One of the P38s was lost, but Yamamoto died in the wreck of his aircraft, and America's most formidable opponent in the Pacific was no more.

Increasingly, American capabilities in reading Japanese ciphers gave them a vital edge in knowing what their opponents were planning to do. Although they remained a formidable threat, the lack of surprise meant that the US Navy was able to gain and keep the upper hand during the remainder of the Solomons campaign. Later, breaking the Japanese merchant-ship code enabled US submarines to cripple enemy convoys and disrupt the enemy's whole resupply and reinforcement system.

During the fighting up the long chain of Pacific atolls, American forces were able to approach islands unseen because intercepted messages revealed the areas that were covered by Japanese early-warning air patrols. Heavily fortified islands were identified and bypassed by the invasion forces, while intercepted messages revealed the huge Japanese air strike aimed at the ships preparing to retake the Mariana Islands. For eight-and-a-half hours, huge swarms of Japanese aircraft attacked the ships of Task Force 58. But the Americans were listening to the radio messages coordinating the strikes, which told them exactly what was coming their way and when. The result became known as "The Great Marianas Turkey Shoot," and the huge losses of trained aircrew all but crippled the Japanese naval air arm.

For the rest of the Pacific fighting, American cryptanalysts were able to help defeat the Imperial Japanese Navy in major engagements like the Battle of Leyte Gulf, during the retaking of the Philippines, and to make it possible to hunt down individual warships that were frantically trying to escape Allied aircraft. Even the last desperate throw of the kamikaze suicide strikes on the U.S. fleet attacking the islands of Iwo Jima and Okinawa was signaled over radio channels monitored by the Americans, providing enough warning to reduce their terrible effect.

Protecting the "Ultra" secrets

OPPOSITE BELOW Lockheed P38 Lightning long-range fighters had the endurance to reach and shoot down Admiral Yamamoto's plane on information derived from secret Japanese ciphers.

BELOW Karl Doenitz, a submarine officer in World War I, was commander of the U-boats, and later of the whole German Navy, in World War II.

For both the British and the Americans, the overriding nightmare was the possibility that in reacting to information gained from reading the ciphers of their German and Japanese adversaries, they would give the game away. If this happened, and their opponents switched to a completely different system, the resulting intelligence blackout would be a major catastrophe. Yet failing to act on vital information that could affect the outcome of a battle or a campaign would be equally disastrous. Thus the whole cipher war became a delicate balancing act, calculated to disguise from the enemy the fact that their most secret messages were being read on a regular basis.

One obvious precaution was to give the information to senior commanders only, and to conceal the actual source of the information under the cover name of "Ultra." Even then, it was difficult to keep the lid on this priceless source of intelligence. The American General Douglas MacArthur was so delighted at the shooting down of Admiral Yamamoto that he wanted to publicize the success all over the world. It took the direct orders of the Joint Chiefs of Staff to keep the secret quiet, and no official references were made to the killing of Yamamoto until the Japanese themselves had announced his death.

In the Battle of the Atlantic, the consequences of reading German signals between the U-boat commander, Admiral Doenitz, and the submarines operating at sea, were more difficult to conceal. Warned of the concentration of U-boats into a wolf pack and given the details of the patrol line they would adopt to search for an approaching Allied convoy, the Allies were able to order ships to change their courses to take them clear of the threat. Because it was never revealed that this was due to precise information, the Germans did not immediately suspect that the security of their machine ciphers had been compromised. On the other hand, because the Germans were reading British naval codes during the early part of the war, they were often able to learn the details of the new convoy courses and instruct their U-boats to move accordingly. These messages would be read, and the convoy course would be changed again, in a macabre sequence of search and evasion across the bleak waters of the North Atlantic.

Later, the Germans became more suspicious and carried out several enquiries to determine whether it was possible for the enemy to have broken into the Enigma system. As in World War I, they were handicapped by their assumption that the security of their systems, and the precautions they had taken to safeguard them, were so formidable that the Allies could not have succeeded in reading their signals, regardless of what events may have suggested.

Even when communications from German prisoners-of-war, using prearranged codes in letters sent back to their families, revealed that their ships had been boarded by Allied sailors, they failed to hear the alarm bells. The regular changes in machine settings, and the huge number of possible relationships between each letter of the message and its Enigma-ciphered equivalent, provided complete reassurance. Added to this, the idea of ranks of brilliant mathematicians and cryptanalysts working with number crunching machines that could try vast numbers of possible relationships until they revealed the correct one, was entirely unknown to the German investigators. Unable to visualize the Allied weapon that could break their security, they concluded that it was perfectly safe.

ABOVE Admiral Isoruku Yamamoto, Commander in Chief of the Imperial Japanese Navy shot down and killed by American fighters in a triumph for cipher analysis.

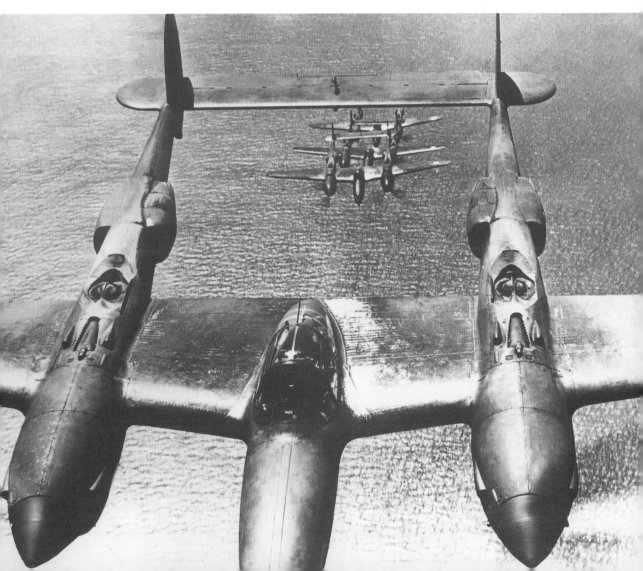

Electronic Secrets – on the Ground:

ABOVE Photographs of a Soviet spy trawler, taken by a Royal Air Force aircraft on a reconnaissance flight over the Atlantic, are examined by an RAF intelligence officer.

OPPOSITE ABOVE AND BELOW Power projection: the American nuclear aircraft carrier *USS George Washington* in the Northern Gulf, as flight-deck crews prepare aircraft for some of the 80–100 sorties flown each day over the "no-fly" zones of Iraq.

In an ideal world, espionage and intelligence would involve strictly one-way traffic. Information would flow inward about an actual or potential opponent's resources, plans, and intentions, but no useful intelligence would pass in the reverse direction to tilt the balance of advantage the other way. In reality, of course, vital information flows in every conceivable direction. Some is channeled through agents, double agents, triple agents, and multiple agents, information that, because of the failings and limitations of human beings, may be valuable intelligence or be deliberately planted as a potentially misleading gift from the opposition's counterintelligence organization. This is the very meat and drink for the very best spy fiction writers, whose realistic backgrounds and convoluted plots hint at the awesome complexities of operating networks of agents in the field.

However, the same two-way flow of information operates in areas of espionage where human loyalties are never involved, in the world of electronic and communications intelligence. By their very nature, systems that are designed to provide information, say on the location of potential targets or the approach of hostile aircraft, tend to betray themselves, their purpose, and their capabilities as soon as they begin to operate. Important radio messages to units of an army or a government, even when protected by complex ciphers, can still reveal much to an electronic eavesdropper, by the level of traffic, the length of the messages, and the locations of the transmitters, without a single message being deciphered successfully.

This is the arcane world of electronic intelligence, where traffic analysis can help to build a detailed picture of an adversary's defenses, his bases, his weaponry, his movements, and his capabilities. Conversely, by very careful protection of these secrets, information revealed through electronic intelligence can be false, incomplete or misleading, so there is an increasingly important need for all possible aspects of the intelligence picture to be studied, to watch for any gaps or inconsistencies. Likewise, the only way a country can avoid giving away potentially valuable information in this way is to totally ignore an apparent threat, or give up the opportunity to train and exercise its defensive forces, a course of inaction that may be far more dangerous than the inevitable release of information that is inseparable from all these activities.

Confrontations on the high seas

During the later Cold War years, the Soviet Union was keen to develop a formidable deep-water navy, which could project Soviet power around the world in a similar manner to their chief adversary, the US Navy. In this superpower confrontation, the Americans had one supreme advantage, in the shape of the carrier battle groups. Each of these was

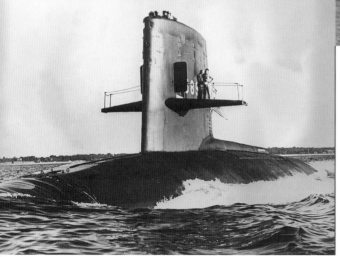

ABOVE A sister ship of the *USS Skipjack*, said to be the world's fastest submarine at the time, *Scorpion* approaches the Royal Navy dockyard at Portsmouth, England, the first nuclear submarine to enter British waters.

LEFT The American nuclear submarine *USS Scorpion* running at speed through Long Island Sound, on her maiden voyage in June 1960.

centered on a nuclear-powered aircraft carrier able to circle the globe over and over again without needing to refuel, and capable of launching a devastating air strike anywhere in the world, with nuclear or conventional weapons, and often beyond the range of land-based air power.

Since originally it was the Americans and British who had developed the basic technology and tactics of carrier warfare, this US Navy capability had evolved from a continuing tradition, reinforced by a vast amount of knowledge, which the Soviets simply did not have. Not only were they faced with having to design carriers from scratch, together with the aircraft that would operate from them, but also they had to learn the basics of carrier operation, which would mean studying how it was done by the experts, at the closest possible distance. At the time, the US Navy was effectively running the only game in town, when it sent its carrier groups to sea to hone their operating skills in cooperation with the navies of their NATO allies.

As a result, no US exercise was complete without the presence of Soviet aircraft, warships, and even fishing trawlers—paramilitary craft loaded with electronics as well as fishing gear—all taking the closest possible interest. At times, these surface ships became a positive threat, using the freedom of the high seas to edge closer to opposing warships than was safe. There were incidents, near misses, even direct collisions as a result of errors, incorrect decisions, and misjudgments of other captains' intentions. The purpose was to gather information: by recording radio transmissions, frequencies, and the locations of transmitters in the traffic between ships and aircraft, much could be learned about how the carrier groups operated. The electronic signatures of radar systems, including the power, the frequency, the pulse rate, and other data, could make them easier to identify and easier to jam or mislead in the event of a real conflict.

Submarine surveillance

However, the surface confrontations, regular and difficult as they were, provided only part of the picture. For sheer danger, there was little to beat the submarine equivalent being fought far below the surface, in the dark abysses of the deepest oceans. Here, the information was even more valuable, and the risks much greater. At stake were not the techniques and weaponry of carrier operations, but the strategy of national survival. At a time when all the major nuclear powers relied on ballistic-missile submarines to deliver the mighty blows on which the concept of deterrence depended, it could be a matter of life or death to keep the closest watch on an adversary's submarines.

Knowing the routes that ballistic-missile submarines would follow to their wartime stations would provide a valuable early warning of hostilities being about to break out. In the main, the World War II expedient of using sound waves to locate submerged submarines is no longer viable for today's submariners, since sending out sonar transmissions reveals as much of the searching submarine's position as it does of its target. Instead, all the hunter-killer submarines can do is listen for the sound signatures of enemy craft in the otherwise silent depths. Often, only the tiniest variations in the beat of the propeller, or the shape of the waveforms of the sound signal, can indicate the difference between the submarine of an ally and that of an adversary. Using this data, a submarine captain may have to decide to launch a nuclear torpedo or take evasive action, and a single error could be fatal for all concerned.

With each side producing new designs of submarine, or modifying existing classes of boat to make them quieter in operation, and thus more difficult to detect, it was essential for every navy to be able to push ever closer to one another's submarines to record and identify the noises they made. The Americans had installed a sound monitoring system called SOSUS (Sound Surveillance System) on the seabed across the deep-water chokepoint of the GIUK (Greenland-Iceland-UK) Gap, through which most Russian submarines passed on their way to the open Atlantic. This could pick up undersea noises over a vast area, but made it even more essential to know and be able to identify the noises made by every class of operational Russian warship.

Deadly accidents

Here too accidents happened—collisions between massive underwater craft maneuvering too close for comfort, or simply disasters that occurred in the everyday routine of operating in the dangerous undersea environment. The sinkings of the *USS Scorpion*, lost in 1968, and more recently the Russians' *Kursk* are thought to have been due to malfunction of the submarines' own weapons, causing catastrophic explosions and the loss of both crews.

Other incidents, often less serious in themselves than in their implications, have been caused by the need for information. In September 1969, the US hunter-killer submarine *Lapon* succeeded in shadowing a Soviet Yankee-class nuclear ballistic-missile submarine through an entire patrol, in an epic of undersea espionage.

But surveillance wasn't always so successful. In the preceding summer, the *USS Tautog* had collided with the Soviet Echo-class submarine she had been shadowing, and while the American boat had survived to return to Pearl Harbor, her crew were convinced that they had heard the other submarine break up and sink. In fact, the Russians later revealed that she too had survived, and the crew had thought that the Americans had perished.

Collisions also happened between British and Soviet submarines. According to American sources, the British nuclear hunter-killer submarine *HMS Sceptre* collided with the Russian nuclear submarine she was shadowing in Arctic waters late in 1981. Later, *HMS Splendid*, of the same class, had her towed-array sonar torn away by a large Russian ballistic-missile submarine in the Barents Sea on Christmas Eve, 1986.

OPPOSITE LEFT The *Kursk* moored at the Vidayevo naval base.

OPPOSITE RIGHT One of the *Kursk*'s officers, Captain-Lieutenant Kolesnikov, and the farewell note he wrote while trapped in the flooded hull after one of the submarine's torpedoes exploded during a test firing in the Barents Sea, blowing a hole in the bows and causing the loss of the vessel and all her crew.

BELOW The crew of the Russian nuclear submarine *Kursk* at a naval review in her home port of Severomorsk.

капитан-лейтенант Колесников :

- "13.15. Весь личный состав из 6,7 и 8 отсеков перешел в 9. Нас здесь 23 человека. Мы приняли это решение в результате аварии. Никто из нас не может подняться на верх ..... Я пишу наощупь".

9

The continuing search for information

Similar confrontations have occurred high in the stratosphere. During the Cold War years, Russian long-range patrol aircraft would approach British airspace by skirting to the north of Norway, and Royal Air Force fighters would be sent up to intercept. Normally, the fighters would take up close formation with the Soviet aircraft and escort it clear of UK territory, but not without giving the Russians unavoidable, but valuable, information. The Soviet airmen would know when they were picked up by British early-warning radar, giving them range, frequency, and location of the ground stations involved. Their own radar would tell them the base from which the fighters came, while the time taken for them to appear alongside would be a measure of their climb performance. The amount of time they remained in formation before being relieved would show their operating endurance.

The search for information would occupy both sides of the Iron Curtain during the height of the Cold War. For years on end, the signals traffic of Warsaw Pact units was recorded and analyzed. Two main facts

resulted from this information – the level of traffic in terms of the number of messages, and the location of each transmitter, as revealed by cross-bearings from direction finders. Over time, particular combinations of callsigns and frequencies would be shown to relate to tank or infantry units, or brigade or divisional headquarters. If traffic levels increased sharply, it tended to indicate either an increase in the state of readiness, or the running of a major exercise. If many of the units moved forward (closer to the border), this reinforced the picture of opposing forces going on to full alert, which was particularly significant at a time of increasing international tension.

Since the breakup of the Warsaw Pact, and the forging of a new relationship between East and West, the pattern of electronic intelligence gathering has changed almost out of recognition. However, the search for information from potential adversaries is still thriving—only those adversaries have changed. British and American flights over Iraq, maintaining the essential knowledge of hostile air defense radars that might be invaluable to the safety of aircrews if shooting should break out, and the loss of the US intelligence aircraft over the China Sea in 2001 (see Chapter 10) are both parts of the same continuing tradition. Knowledge is power, and never more so than in the complex and competitive world of military electronic intelligence.

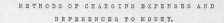

If the ideal spy is one who works loyally and unobtrusively in enemy territory to collect vital information, a double agent must be the next best thing. Because a double agent is often highly placed in their own country's defense, government or intelligence organization, the information produced is often very valuable. On the other hand, the change in allegiance that creates a double agent can produce an inherently unstable relationship. There is always the danger that an apparent double agent is actually a deliberate plant from the opposition's intelligence services, and the valuable information they deliver may be laced with disinformation to compromise its worth. Even when a double agent is genuine enough, fear or greed, or residual loyalty to their own country can persuade them to switch sides again, to become a treble or even multiple agent, whereupon their real worth and that of the information they deliver becomes very dubious.

One of the most complex agents in the history of espionage was Ignatz, or Isaac Trebitsch, born into a Hungarian Jewish family in 1879. After giving up rabbinical studies, he moved to London at the age of 17 and joined the Church of England. Then he went to Germany and on to Canada, where he was ordained as

```
METHODS OF CHARGING EXPENSES AND
        REFERENCES TO MONEY.
-----------------------------------------------

Letter No. 10.  31.10.41.

        (a)  Personal expenses.  Hotel, tramways, etc.,
             at £2 per day.

        (b)  Agents at £2 per day each.

        (c)  Hire of typewriter.

        (d)  Post and paper

        (e)  English books.

        (f)  English lessons.

        (g)  Drinks and taxis.

Letter No. 11. 3.11.41.

        Detailed list of cost of all agents' journeys.

Letter No. 15. 24.11.41.

        I need a reserve fund of money here.

Letter No. 29. 7.2.42.

        (a)  Am only given £10 bonuses to cut down expenses.
             (Given each month)

        (b)  Travelling expenses are paid to them apart.

Letter No. 52. 30.5.42.

        Money can be sent to the National Provincial Bank,
        Swiss Cottage Branch, in name of Juan P. GARCIA,
        as representing payments of commissions.

Letter No. 65. 12.7.42.

        If you don't consider my plan for sending money to the
        Bank a good one you must study alternative means and
        let me know.

Cover letter 12.7.42.

        (a)  I shall need about £200 per month for self, wife
             and child.

        (b)  This does not include incidental expenses.
```

ABOVE German documents relating to expenses payments for their agent "Cato," who was actually one of the Allies' most successful double agents.

LEFT Aldrich Ames and his wife are escorted by a federal marshal from the courthouse in Alexandria, Virginia, while being tried on charges of spying for the Soviet Union. During the hearing, US District Judge Claude Hilton extended the order freezing some $2.2 million of assets allegedly belonging to Ames.

RIGHT AND FAR RIGHT Juan Pujol Garcia volunteered to serve as an Allied double agent, and subsequently was recruited by the Germans under their codename "Cato." Known to the Allies as "Garbo," he fed the Abwehr a diet of carefully prepared Allied disinformation, which played a vital part in the D-Day deceptions.

an Anglican minister in Montreal, at the age of 23. Seven years later, he returned to London and changed his surname to Lincoln, taking British nationality so that he could stand for election as a member of Parliament. He won a seat for the Liberal Party in 1910, but money problems forced him to develop a career as a freelance journalist. He went to report on the Balkan Wars in 1912, and became a double agent for Bulgaria and Turkey, although the speed with which the Germans had him released from a Bulgarian jail when his activities were discovered made it almost inevitable that he had been working for Berlin as well.

Trebitsch returned to London and tried to join British Intelligence. When they turned him down, he moved to the USA, but was extradited to the UK for fraud, and jailed, before being deported to his native Hungary in 1919. From there, he traveled back to Germany, where he was involved in an unsuccessful coup in 1920, after which he moved to China. The intelligence service of the Chinese Nationalists employed him as an agent against the independent warlords, although he was rumored to have been working as a double agent for the Japanese. He became a convert to Buddhism, and by the time he died in 1943, in Shanghai, he had become an abbot with the Chinese name of Chao Kung.

Not all double agents have such a complex and nomadic life. William Sebold was a German-born American who was blackmailed by the Gestapo, while visiting his mother in Germany in 1939, and forced into joining the Abwehr. He alerted the American authorities who arranged for him to work in the US as a double agent, tipping them off about German spy networks in

America and sending false information back to his controllers in Hamburg. He testified as a witness in the trials of German spies captured in America, then returned to civilian life after the war.

Oleg Gordievsky, a double agent for the British since 1974, ultimately became the KGB resident at the Soviet Union's London embassy, able to provide information on the rising star Mikhail Gorbachev, which proved invaluable when British Prime Minister Margaret Thatcher and US President Ronald Reagan negotiated with the newly appointed Soviet leader. He also gave the West a long list of names of pro-Soviet agents, but was recalled to Moscow after being betrayed by American double agent Aldrich Ames in 1985. Because Ames worked in the CIA's Directorate of Operations, he was also able to betray a whole network of Western agents, but the British managed to smuggle Gordievsky across the Soviet border to safety in a diplomatic van from their Moscow embassy.

Probably the most successful double agent of all time, in terms of the effects of his information on the course of history, was a young Spaniard named Juan Pujol. He volunteered to work for British Intelligence, but was turned down. Then he contacted the German Abwehr and was taken on by them as an agent, being sent to the UK, where he contacted the British again. This time he was taken on as a double agent. Under his codename of Garbo, he played a vital role in feeding the Germans disinformation in the Fortitude D-Day deceptions campaign, which persuaded them to keep massive forces tied up waiting for an attack that never came, while the Normandy beachhead was firmly established (see Chapter 7).

Listening to the enemy

TOP Commander of the Russian Second Army at the Battle of Tannenberg, General Alexander Samsonov shot himself on August 29, 1914 while fleeing from the advancing Germans.

ABOVE Former cavalry officer Paul von Rennenkampf commanded the Russian First Army and sent orders by radio, which were intercepted by the Germans.

From the earliest days of World War I, Imperial Germany was faced with having to fight on two fronts: the Western, against the French, British, and Belgian Armies; and the Eastern, against the forces of Tsarist Russia. The idea of the "Russian steamroller," vast masses of peasant soldiers pouring unchecked over the eastern border had been the cause of German nightmares for generations. However, the Schlieffen master plan had assumed that the Russians would need time to assemble sufficient forces to be a serious threat, and in the meantime the bulk of the German Army could be used to force a decision in the west, after which it could be sent east to beat the Russians.

In fact, the Russians moved much more quickly, and effectively, than anyone expected. Within days of the outbreak of hostilities, and a month before the Germans predicted, they had invaded Galicia, a province of the Austro-Hungarian Empire located in what is now Poland. They had also sent two armies onto the sacred soil of Prussia, one of which managed to defeat a German force twice its size at Gumbinnen. This dreadful shock led to the dismissal of the German commander. He was replaced by a retired general, Paul von Hindenburg, while General Erich Ludendorff was rushed from the fighting in Belgium to serve as his chief of staff.

They faced a serious situation. To the north was the First Russian Army under General Pavel Rennenkampf, advancing from Gumbinnen, and to the south the Second, under General Alexander Samsonov, close to the village of Tannenberg. If these two armies were able to join forces, the already demoralized Germans could scarcely hope to defeat them, but they had one trump card in their hand. In their extreme haste to mobilize and attack Germany in support of their French allies, the Russians had not had time to organize properly. Communications depended on radio, but the two armies had been issued with different ciphers, so they could only talk to each other in plain language. An astute German staff officer, Colonel Hoffmann, had intercepted these messages and knew that he could eavesdrop on Russian plans at the highest level.

This information was literally priceless. Hindenburg and Ludendorff wanted to attack Samsonov at Tannenberg, before Rennenkampf could march to his aid. On August 25, 1914, German troops hurriedly withdrawn from the north moved to attack Samsonov's army. Hoffmann, on his way to the scene, stopped his car at the railroad station at Montovo to call his headquarters. He was told that two Russian radio messages

had been intercepted that morning. The first of them was crucial; it contained Rennenkampf's orders to his army to march south. That much was predicted, and worrying, but what made the message so valuable was its estimate of the distances to be covered the next day. Because his ponderous and ill-equipped army would not be close enough to threaten the Germans pounding Samsonov, it was safe for the German attack to go ahead.

Hoffmann jumped into his car and roared off in pursuit of Hindenburg and Ludendorff. Within a few miles, he had caught them up, and as his car drew alongside, his driver passed the intercepted messages across. The result was that on the following day, August 26, 1914, Samsonov's army was outmaneuvered. The Russians, believing the Germans to be in retreat, were actually advancing when their better-informed enemy launched an onslaught. The Russians were thrown back and encircled, and by the end of the following two days they were in headlong retreat.

From their reading of Rennenkampf's radio messages, the Germans knew that he was still advancing, but west rather than south, and there was no danger of him reaching the battle area. By the time he turned southward to aid his comrades, the Germans were ready for him. Two more army corps were being rushed by train from the Western Front to reinforce the defenders of East Prussia, and the Tsar's armies never again posed a serious threat to German territory.

Stalking submarines

One of the most successful American espionage submarines during the late 1960s was the hunter-killer *USS Lapon*. In a series of epic voyages, she excelled at the dangerous task of snooping on the growing Soviet nuclear submarine force, displaying ever greater daring. Her successes began in March 1969, when patrolling through the icy waters of the Barents Sea. At the entrance to the Soviet Union's premier Arctic submarine base, her skipper, Captain Chester M. Mack, spotted the awesome sight of a Russian Yankee-class nuclear ballistic-missile submarine through his periscope. This was a major scoop—the first clear sighting of the Soviet equivalent of the American Polaris missile submarines—and Mack was able to take a series of photographs through the periscope to give Western intelligence their clearest views yet of this powerful new threat.

Yet this was only the beginning. Knowing what the Yankees looked like was useful enough, but what was really needed was information on their performance, their maneuverability, and their audio signatures, so that anyone passing in sonar range of one would recognize the sounds they heard. There was only one sure way to gain this information, and that was to tail one of the Yankees for as long as possible on an operational patrol, without the Russian crew being aware of the Americans' presence.

The *Lapon* and her crew had already come too close for comfort to Russian submarine and anti-submarine forces. On one patrol in the Barents Sea, her commander had looked through the periscope to find himself being watched by a Soviet anti-submarine helicopter, which had spotted the "feather," the small wake trailed behind the periscope as it broke the surface. On another, she had been watching two Russian submarines conducting practice attack drills, when her sonar revealed torpedoes in the water running toward them. Taking evasive action at full speed, she avoided the practice torpedoes which, although they almost certainly carried no warheads, gave everyone on board a very anxious time.

Trailing a Yankee, though, would require very careful planning. A few months before the first sighting, the *USS Greenling* had managed to slip beneath the hull of another Yankee to record some of the sound emissions from the Russian boat, but in audio terms, this was merely a snapshot rather than a full-length feature. Mack had arranged to be sent a signal when the SOSUS underwater sonar surveillance system picked up a trace of a Yankee heading for the Greenland-Iceland-UK Gap, using the *Greenling* tapes for identification. The call came through on September 16, and as the *Lapon* headed for the area, a US Navy Orion anti-submarine patrol plane confirmed the Soviet submarine's heading.

The American submarine reached her target destination, off the Denmark Strait between Iceland and Greenland, a day before the Russian showed up. There followed a frustrating game of hide and seek that lasted for four days, during which the Yankee's sounds would often be drowned out by the clamor of marine life and nearby fishing boats. After that, the Soviet boat turned on a steady course, allowing the *Lapon* to follow close behind. Then, after 18 hours, she made an unpredicted turn and disappeared. No sound trace remained, and the audio scent was lost.

Mack decided on a cold-blooded gamble: he would move ahead of the Russian submarine and attempt to predict her destination. He sailed down to the Azores, but his search was unsuccessful. With a sinking heart, he decided to head west toward the distant coast of the USA, but ran into a trawler's nets. The crew of the fishing boat cut away the gear, but the *Lapon* was left with a cable trapped across her bows, creating a thunderous underwater turbulence that would enable the Russians to hear her from miles away.

The submarine was forced to surface, itself a noisy maneuver, so that the cable could be cut away. After diving again and waiting another day, their patience was rewarded. The faint sounds of the approaching Yankee were heard by the *Lapon* 's sonar team. This time, Mack was determined not to lose the Russian, and he closed to within half a mile of his quarry. Because the Yankee was quietest approached from astern, and noisiest on her port beam, Mack had to try to keep on the Russians' port side. For the rest of the voyage, the *Lapon* hung on grimly, trying to match every twist and turn of the Soviet vessel, while the crew recorded the noises she made and attempted to avoid emitting any unnecessary sounds themselves. Even a slammed compartment door or the dropping of a wrench would have alerted the Russians to the presence of a shadowing vessel.

In all, the *Lapon* remained in position for no fewer than 47 days. During that time, the crew were able to map out the Yankee's operational area and track her through regular sharp turns, including several that traced a complete circle. At one stage, the Yankee turned around and headed straight at the American submarine, which had to dive deeper to stay clear. This maneuver, called a "Crazy Ivan" by the US Navy, was intended to reveal any shadowing craft in the vicinity, but the noise made while carrying out the turn and making the high-speed dash deafened the Soviet sonars, and the Americans remained undiscovered. Finally, the Yankee turned for home at the end of her patrol, steering a straight course to her own coastal waters. The *Lapon*, duty done, was able to go home too.

OPPOSITE TOP The launching of the US Navy hunter-killer nuclear submarine *USS Lapon*.

BACKGROUNDS Nuclear submarines can dive suddenly and almost without trace to seek the protection of the depths, where pursuit is more difficult.

Tunnel vision

Not all espionage coups involve undercover agents penetrating enemy territory to collect information and returning across the border to report to their controllers. When the United States commissioned a new purpose-built embassy in Moscow, in 1953, the KGB had already moved in alongside the builders, placing more than 40 hidden microphones in the structure to relay the most secret conversations to Soviet ears. Even an official gift, the Great Seal of the United States, which hung on the wall behind the ambassador's desk, was home to a hidden mike, and much damage was done before the bugs were discovered.

On the other hand, Western intelligence services have been able to eavesdrop on the conversations of their Cold War adversaries on at least two occasions, by tunneling into their communications systems and listening in to secret messages without a single spy being put at risk. And in one case, the other

American sector

Russian sector

British sector

side was tipped off by one of their own agents, yet they dared take no action because of the danger of exposing the presence of their man in the enemy camp.

The first tunnel coup occurred in 1951, when Austria – like Germany – was still occupied by the four Allied powers: Russia, the United States, Britain, and France. Each nation administered a different zone of the country, but the capital, Vienna – like Berlin – was split between the four powers. This had two implications for the intelligence services of the occupying powers. Because Vienna was a compact city, all four Allied headquarters were close to one another in the center. Furthermore, all the telephone networks for the whole country were routed through Vienna, in a complex web that crossed back and forth between the different zones of the city.

This situation allowed the British Secret Intelligence Service to dig a tunnel beneath the foundations of their Vienna headquarters, which enabled them to tap into the telephone lines that ran to the Soviet headquarters in the Imperial Hotel, on Vienna's Ringstrasse. Some of the information revealed when they broke into the Russian signals network was in clear, and could be read with ease. The most important information, however, was in code, and this was much more difficult to translate, until their American opposite numbers in the CIA came up with a key.

The CIA had already been doing their own mapping of the underground cable routes, when the British told them of their tunnel, codenamed Operation Silver. Ironically, it was a defect with one of the CIA's own cipher machines that provided the means to overcome the Russian codes. Tests had shown that when the operator typed the clear text into the machine, as the German operators had done with the Enigma (see Chapter 5), the process had sent a faint signal revealing the plain text, which could travel for miles along the cables. This caused the CIA machine to be scrapped, but knowledge of the problem made it worth investigating whether or not the Russian cipher machines produced the same effect.

Sure enough, they did. For several years, the overjoyed intelligence men were able to learn all manner of sensitive information, including the Russians' reluctance to invade Yugoslavia during Stalin's dispute with Tito, which proved invaluable in Cold War power-politics.

The success of Operation Silver led to the even more spectacular Operation Gold, in Occupied Berlin. Using information supplied by the German spymaster Reinhard Gehlen, once again the SIS and the CIA collaborated in a plan to dig a tunnel from Western territory. This started

OPPOSITE TOP General Reinhard Gehlen, former head of the West German Intelligence Agency.

OPPOSITE BOTTOM Focus of espionage: Potsdammer Platz, the meeting point in Berlin of the Soviet, American, and British sectors, indicated by the black lines. This photograph was taken in August 1948 (inset shows the four Allied sectors in post-war Vienna).

BELOW A police officer guarding the border between the British and American sectors in Berlin, February 1948.

YOU ARE NOW LEAVING BRITISH SECTOR

ABOVE The Russian spy George Blake betrayed the SIS/CIA tunnel in Berlin to his Soviet spymasters, but eventually was caught and sentenced to 42 years in prison.

OPPOSITE The 350-yard tunnel extended from the American sector of Berlin beneath the Soviet sector.

OPPOSITE TOP After Blake's warning, a Russian soldier examines the inside of the tunnel to check for telephone tapping equipment.

OPPOSITE BOTTOM LEFT George Blake, seen with his mother in Russia during the late 1960s, escaped from Wormwood Scrubs prison in London in 1967.

OPPOSITE BOTTOM RIGHT West German Intelligence chief General Reinhard Gehlen.

in a cemetery in the suburb of Alt Glienecke, in the American zone, and ran a distance of 500 yards, at a depth of 15 feet, under the inter-zonal border to tap into the Russian landlines. These ran from their Berlin headquarters in Karlshorst, on the south-eastern side of the city, to Leipzig in the east, then on to other destinations in eastern Europe and the Soviet Union itself.

It was a formidable undertaking, and it should have been doomed from the outset, since the Russian double agent George Blake had known about Operation Gold while it was still being planned. Nevertheless, the Russians were totally unaware of the technical weakness of their encoding machines, so they decided to let the Allied operation continue, rather than risk revealing that they had a "mole" in place on the Western side. With all their sensitive signals protected by unbreakable codes, this seemed a sensible enough precaution.

The tunnel was completed by US Army engineers on February 25, 1955. The earth dug out during construction was disposed of carefully, while air conditioning was installed to prevent the ground above the tunnel from warming up and revealing its course. The CIA began taping messages carried on three landline networks, each comprising a telegraph cable and four telephone cables. With each line carrying multiple conversations simultaneously, the equipment installed at the head of the tunnel was recording up to 1,200 hours of material daily on 600 tape recorders, producing 800 tapes a day, which were flown out once a week to London and Washington.

Operation Gold ran for just over a year. In the spring of 1956, heavy rain had caused problems with water seeping into telephone cables in East Berlin, and when, on April 21, the East German engineers started digging up the cables to carry out repairs, they were horrified to discover the tunnel, which their Russian masters had kept a secret. The information channel was cut off, but for 14 months the Western allies had read information passing between the Soviet and East German governments, the Russian embassy in Berlin, and the Soviet Army headquarters at Zossen, outside Berlin, revealing vital information on the Soviet forces in eastern Germany.

LEFT The underground operations center for the Berlin tunnel.

Listening to the Soviets

After the end of World War II, when the reality of the Cold War became all too apparent, an early Western priority was to unravel as much as possible of the Soviet Union's communications and defense organizations. Although spies were still extremely valuable, it was becoming exceedingly difficult even for a highly-placed double agent to gain access to more than a small part of the intelligence picture. Consequently, much of the espionage effort had to be sustained by a combination of ground monitoring stations and airborne surveillance, particularly after the Western radars supplied to Russia under wartime agreements were progressively replaced by their own designs.

For the USA, much of this effort had to be made from the territory of allies rather than American soil, because of the ranges involved. In 1952, two surveillance sites were established in the United Kingdom—one at a village called Kirknewton, near Edinburgh, which began recording Russian teleprinter messages, and the other at a wartime radio-intercept station at Chicksands Priory, which listened to conversations between Soviet pilots and their ground controllers. Stations were also set up in Turkey to monitor Soviet naval operations in the Black Sea, and missile tests on nearby ranges. By the middle of 1955, long-range American radar at Samsun, on Turkey's Black Sea coast, was monitoring the test flights of Russian ballistic missiles over ranges in central Asia and across the Russian far east into the Pacific, near Vladivostok. The information this system provided gave the Americans details of the missiles' speed, track, altitude and range, and also, from the timetables of the test flights, the time when each new missile went into quantity production.

Additional information was provided by flights close to and, in some cases, across the Russian borders by both the British and the Americans. US flights over Europe began from British bases, like Sculthorpe and Lakenheath in Norfolk, Manston in Kent, and Brize Norton in Oxfordshire. Bases in Japan and the Western Pacific were used to cover the other end of the Soviet Union's huge land mass. Several of the aircraft were shot down over Soviet territory, but in at least one case, on July 29, 1953, an American RB50 reconnaissance plane was shot down by Russian fighters in international airspace, over the Sea of Japan. Sometimes these attacks were made on non-intelligence-gathering aircraft by mistake: on June 22, 1955, the Russians shot down a C118 transport, which had strayed over the southern Soviet Union following a navigational error. In this case, the plane made a crash landing and was destroyed on the ground by its crew, who were returned to the US after prolonged questioning.

ABOVE No tactic was too underhand, and no equipment or ploy too simple for either side once the Cold War began in earnest. Phone tapping was one of the most widely used, and discovered, methods of espionage.

NATO Exercise Able Archer

Russian fears of a more belligerent NATO were fanned by American interest in the predicted test flight of the Soviet SS-X-24 multiple-warhead missile on 31 August, 1983. The missile was due to hit the test range on the Kamchatka Peninsula, just west of Japan, and as part of an intelligence gathering mission, the US had sent an RC135 reconnaissance aircraft to make photographic and electronic recordings of the results. By a deadly coincidence, a navigational error caused a Korean airliner to wander across Kamchatka at the same time. The Russians mistook it for an American spyplane and shot it down, with the loss of all 269 on board (see Chapter 8).

The West reacted furiously to the Russian attack on a civilian aircraft, further intensifying Soviet paranoia. Two months later, NATO began a major exercise called Able Archer, which could not have been more provocative from a Russian viewpoint. Although not a single warship, aircraft nor soldier was involved, the exercise simulated the orders that would be given to intensify

ABOVE Part of a seat label with script in English and Korean, from the Korean Air Lines Boeing 747, recovered from the sea off the Russian coast.

BELOW Soviet Sukhoi 15 fighters are scrambled to intercept the 747 over Russian airspace; one of them is ordered to shoot it down with an air-to-air missile.

ABOVE Oleg Gordievsky, double agent and former KGB chief in London, warned British Intelligence about alleged contacts between Soviet spies and former British Labour Party leader Michael Foot, who denied the accusations.

BELOW Japanese fishermen searching for debris from the Korean airliner shot down by Soviet fighters.

levels of readiness and finally to launch a nuclear strike in response to a Russian attack. Because of the sensitivity of the subject, communications routines were changed, and some bases maintained strict radio silence. Since these were two of the changes that the Russians thought would indicate a genuine threat, they assumed that these, together with the huge increase in signals traffic, represented a real danger.

During the course of the exercise, NATO intelligence gathering stations monitored the actions of their Warsaw Pact adversaries. They noted an unmistakable upsurge in signals traffic between military units on the other side of the Iron Curtain. Several nuclear-strike air force units in East Germany were put on alert, although this was not known until double agent Oleg Gordievsky revealed the fact to his Western controllers. Following the conclusion of Able Archer, the first American land-based cruise missiles were installed in Western Europe, and Russian anxiety levels went off the scale. Yet within a year, two of the most prominent alarmists in the highest echelons of Soviet power were gone, following the dismissal of the chief of staff and the death of the defense minister, and East-West relations sank back quickly to their normal level of uneasy coexistence.

ABOVE Korean Air Lines' chief accident investigator, Suk Jin Gu, examines fragments of the wreckage retrieved from the sea off the coast of Wakkanai.

Operation Ryan

By 1980, relations between the two superpowers were becoming chilly. The Soviet invasion of Afghanistan, at the end of the previous year, had provoked US reactions, ranging from a refusal to ratify the second Strategic Arms Limitation Treaty (SALT II) and a boycott of the 1980 Olympic Games to the election of a president, Ronald Reagan, who was seen by the Soviets as very much of a hawk where foreign relations were concerned.

By 1981, the Soviet leader, Leonid Brezhnev, was convinced that the US and NATO were preparing for war, and the KGB was ordered to find out as much as possible about these moves, as part of an operation codenamed RYAN (for Raketno Yadernoye Napdenie, or Nuclear Missile Attack). This required the use of the full battery of Soviet communications surveillance technology, including Cosmos satellites and low-orbit spacecraft that recorded radar signals, and ground stations in Russia, eastern Europe, and most importantly at Lourdes in Cuba, which was ideally placed to eavesdrop on messages between America and her European allies, and US forces overseas.

But perhaps the most unexpected of these listening posts were located in the heart of US territory, within the metropolitan north-eastern states. There were three buildings in the Washington diplomatic enclave that were festooned with communications gear, equally able to read American signals as to transmit their own. These were the old and new Soviet embassies, and a building that served as the headquarters for the Soviet military attaché. There were four sites in New York City—one located within the Soviet mission to the United Nations, another at the Soviet diplomatic residential complex in Riverdale, and two at Soviet diplomatic

BACKGROUND Russian troops in Afganistan.

BELOW Leonid Brezhnev, President of the Soviet Union and General Secretary of the Communist Party (center) with Soviet Defense Minister Marshal Dmitri Ustinov (front row, left) at a parade in Kiev.

recreational facilities on Long Island. Other listening posts were maintained at another recreational facility at Pioneer Point in Maryland, and farther west at Soviet consular offices in both Chicago and San Francisco.

All of these monitoring stations were able to intercept telephone conversations in clear between official limousines and government offices at the State Department, the Pentagon, and the CIA headquarters. The Maryland post could also pick up signals to and from the Atlantic Fleet headquarters at Norfolk, Virginia, and the Tactical Air Command headquarters at Langley Air Force Base. On the European side of the Atlantic, similar eavesdropping was conducted from four diplomatic sites in the UK, seven in Western Germany, two in France, four in the Netherlands, two in Norway, and two in Italy.

What kind of information were they looking for? Targets included any signs of increasing readiness for war, which they identified as upsurges in propaganda offensives against the Eastern bloc, the sending of sabotage teams to Warsaw Pact territory, and a crackdown against Communist sympathizers in the West. Unfortunately, the common tendency for intelligence organizations to see what they expect to see was borne out by several events at the time. On March 23, 1983, President Reagan made his famous "evil empire" speech, denouncing the Soviet Union's repression and belligerence, and made the first reference to the "Star Wars" anti-ballistic-missile defense system. On June 9, Reagan's staunchest ally, Margaret Thatcher, was re-elected as British prime minister, which added to Soviet paranoia. The stage was set for an event to trigger an explosive reaction.

ABOVE Ronald Reagan takes the oath of office as the 40th President of the United States, administered by Chief Justice Warren Burger. Nancy Reagan and Senator Mark Hatfield look on.

BELOW An aerial view of the Pentagon, for 50 years the military nerve center of the USA and the world's largest office building.

CHAPTER 7

Deliberate Deceptions

ABOVE Monsieur le Maréchal de Tallard, Marlborough's vanquished and outsmarted adversary.

OPPOSITE The Battle of Blenheim, where Marlborough and his German allies defeated the French on August 13, 1704.

Not all spy campaigns are straightforward. The history of espionage is riddled with failures and disasters resulting from inadequate training, human error, and pure bad luck. The real problem arises when an intelligence service's agents are captured by the other side, especially if their seizure is not known to their controllers. For captured agents can be turned, transforming the enemy's disadvantage into a powerful, and possibly deadly, advantage. Before the spy's controllers have had sufficient time to become suspicious through a failure to report at a particular time, the agent can be sending false information, carefully doctored to conceal the adversary's actions, plans, or capabilities.

Because intelligence services are always aware of the possibility of agents being captured, becoming double agents, and supplying false information, planted messages have to be created very carefully indeed. Imagine, for a moment, that a KGB agent has been captured by, or defected to, the West. The agent's messages continue, so at first the KGB has no real cause for suspicion. But if the information is designed to serve the West's objectives without giving away too much to the Russians, it will usually need a firm framework of truth, since the KGB will compare it with the contents of earlier messages; a sudden change in part of the intelligence picture would call everything into question.

If a message contains facts that can be checked independently, the KGB is almost certain to do so to ascertain whether the entire message is likely to be genuine. If it is particularly important to convince the KGB of the truth of a message, the double agent's controllers may even include sensitive material that the KGB would never expect to receive in

a misleading report, reinforcing the impression that the message must be genuine. However, the potential result must justify the price of losing that sensitive information.

As with so much else in the world of espionage, the employment of false information has a long pedigree. The Chinese master of the art of war, Sun Tzu, laid down the principles more than 25 centuries ago. He taught that the successful commander should "when capable, feign incapacity, when active, inactivity" and that "when near, make it appear that you are far away; when far away, that you are near." In military terms, he advised the leader to "offer the enemy a bait to lure him; feign disorder and strike him."

The Duke of Marlborough steals a march on his adversaries

In the centuries that followed Sun Tzu's teachings, great commanders made an art form of employing military deceptions. One of the masters was John Churchill, Duke of Marlborough, in his campaigns alongside the Dutch and Germans against the French and Bavarians, almost 300 years ago. In the spring of 1704, Marlborough led his army to the Rhine, while his opponents, under the French Marshal Tallard, anxiously waited to see where his next blow would fall. At Coblenz, Marlborough's army crossed to the opposite bank of the Rhine on a bridge of boats, then marched on upriver. Tallard commissioned a network of spies to discover Marlborough's plans. Eventually, one of them brought the news he had been waiting for, describing the fatal mistake the British had made, which would enable Tallard to defeat them.

Tallard's agent reported that British engineers were building another bridge of boats at Philippsburg, 120 miles farther up the river, which would allow Marlborough's army to cross back to the western bank. This could only mean an attack on Alsace, so immediately Tallard issued orders for his troops to march south, where they would lie in wait for the British and defeat them with a surprise attack.

Unfortunately, the bridge building was a deliberate piece of misinformation. While Tallard's army marched south, Marlborough led his force away from the river to the east, and 30 days after leaving Coblenz, he joined his German allies on the Danube, 500 miles from where the French were waiting in vain. When Tallard realized that he had been outsmarted, he set off in pursuit, driving his troops in a series of forced marches. He found the British and Germans waiting for him, near a Danube village named Blenheim, but by then his men were tired and confused, and in no state to face a determined foe.

The result was one of Marlborough's greatest victories, leading to the destruction of the French Army, the capture of Marshal Tallard, and the routing of their Bavarian allies. Marlborough's careful policy of disinformation had kept the powerful French Army tied down in the wrong place for almost three months, then it had been tempted to its final and catastrophic defeat.

Other commanders have been equally successful in misleading and confusing their

opponents through laying a trail of false intelligence, made to look sufficiently convincing for its finders to accept it as the truth. George Washington was able to conceal his relative weakness from the British on several occasions during the American War of Independence by leaving papers for his enemies to capture, suggesting that his strength was much greater than it actually was. He concealed his casualties by having them buried at night, in unmarked graves, and was able to evade pursuit by threatening to attack.

Disinformation is as valuable in peacetime espionage as it is in war. But the greater stakes and clearer results of wartime deception plans provide the most useful lessons. In many campaigns in both world wars for example, the simplest of materials, allied to the sharpest ingenuity, produced results out of all

proportion to the means employed. Very often, as was the case with Marlborough's campaign against the French, the purpose of false information is to deceive an opponent as to where a decisive blow will fall. And although the means of disguising the falseness of the information, and of delivering it into the enemy's hands, may have changed, the purpose of the coup and its ultimate value have not.

TOP LEFT Dummy horses designed to convince the Turks that the British attack would be delivered against their Gaza positions.

TOP RIGHT Dummy tanks being assembled in a British workshop.

BELOW Dummy soldiers manning a trench position to mislead the enemy as to the whereabouts of the main mass of British infantry.

Deceiving the Turks

In the campaign against the Turks in Palestine
in World War I, the British General Allenby
faced a strong Turkish defensive position at
Gaza. The classic remedy was to outflank this
position by moving his forces east and
attacking Beersheba, but this would only work
if the Turks failed to suspect what was on his
mind. There was always the danger that
movements of troops toward Beersheba might
be spotted by Turkish reconnaissance, which
would reveal what was happening. Instead, it
was decided to deceive the Turks into thinking
that any movements in that direction were a
false trail, designed to divert their attention
from the main blow, which would be delivered
against Gaza.

To prevent the Turks from rejecting the
idea as false information planted by the
British, they had to be made to find the
details for themselves. Colonel Richard
Meinertzhagen, an officer on Allenby's staff,
set off into "no man's land" in front of the
Turkish positions until he encountered
a Turkish patrol. They started firing at him,
whereupon he fled, dropping a water bottle, a
haversack, and a pair of binoculars in his
haste to escape. The haversack had been
prepared very carefully before he set out, and
was partly soaked in horse's blood to imply
that Meinertzhagen had been wounded by
Turkish bullets and had dropped it in his
weakened condition.

Inside were some papers, some money, a
cipher book, and a letter from the officer's wife,
none of which would have been left behind
willingly. Among the papers were some that
referred to a planned attack on Beersheba,
together with the original date for the attack,
which might well have been revealed by the
Turks' own spies. The papers also disclosed
that the plan had been changed to a later
attack on Gaza, which would be preceded by
threatening moves designed to convince the
Turks that Beersheba was the target.

ABOVE The outcome of the successful deceptions: after
driving the Turks out of Jerusalem, British troops guard
the Jaffa Gate.

This plan was based on a sufficiently truthful
framework to fit in with whatever half-picture
Turkish intelligence had been able to gain of
British intentions. Later, messages were radioed
in the cipher that the Turks had captured,
referring to Meinertzhagen having been court-
martialled for losing the haversack. The cavalry,
which would form the main body for the attack,
were moved toward Beersheba in secret, leaving
dummy horses made of straw in their original
positions. The deception was backed up by
heavy radio traffic in the Gaza area, and relative
silence around Beersheba.

The result was a triumph that underlined
the priceless worth of disinformation, properly
applied. The attack at Beersheba, beginning

with a devastating bombardment from carefully hidden artillery, followed by a cavalry charge, achieved such complete surprise that the defenders set off in full retreat. The panic spread to the Gaza positions, where a follow-up attack took these defenses too. The Turks were in full retreat, and within six weeks, Allenby's army had taken Jerusalem.

Falsehoods by radio

Similar ploys were planned and carried out in World War II. During their victorious campaign in France, the Germans used several "black" radio stations, which pretended to be French transmitters conveying essential information to the population. They emphasized how far and how fast the German forces were advancing, and advised people to cash in their savings, causing a run on the banks and floods of refugees to jam the roads. They also fostered the myth of the "Fifth Columnists," German agents parachuted behind the lines in French uniforms to give false and contradictory orders, and hamper the defenses.

This was a brilliant plan, since it caused people to question essential orders given by genuine French soldiers and civilian officials, which tied up the defenders in an even tighter web of confusion. Later, the same tactic was tried against Britain, by dropping hundreds of empty parachutes, each hinting at a spy or Fifth Columnist who had been infiltrated successfully. The apparent threat was met by the formation of the Home Guard, whose primary purpose was to deal with this non-existent force.

Very often, given the capability of modern reconnaissance aircraft, it was essential for false information to be reinforced by evidence on the ground. In the North African campaign during World War II, the German Afrika Korps had pushed the British Eighth Army most of the way back toward Cairo and the Suez Canal by the late summer of 1942. But close to a small station on the coastal railroad, named El Alamein, the wide open spaces of the desert, which had been so well suited to the German commander General Rommel's tactics, gave way to a restricted 40-mile gap between the sea in the north and the impassable salt marshes of the Qattara Depression in the south. Here, the Eighth Army made its stand and finally halted the German advance.

ABOVE A patrol of the British Home Guard, the citizens' volunteer militia raised to guard against a German invasion of Britain in 1940, arrests suspected enemy agents during an exercise.

OPPOSITE BELOW Troops of the Home Guard demonstrate different camouflage and concealment methods designed to give them the advantage of surprise if the enemy should appear.

Blocking Rommel's advance on Cairo

The British managed to change the balance of power on this crucial battlefield through a classic campaign of false information. First, they had to stop Rommel in his tracks by persuading him to attack the strongest point in the British defenses, the ridge at Alam Halfa, which was protected in front by a tract of soft sand that would cause tanks to bog down, and from behind by a line of hidden anti-tank guns. Knowing that the Germans were short of good, accurate desert maps, to the point where captured British maps were highly prized, the Eighth Army's intelligence experts created a series of false documents that were planted on the enemy in several different ways.

These maps showed that the sand was soft and treacherous to the north-east, where the Germans might otherwise have aimed the attack for Cairo. On the other hand, they indicated that there was good hard sand, capable of supporting armored vehicles, leading up to the Alam Halfa ridge. If the Germans accepted the maps as genuine, they were almost certain to fall into the trap. To ensure that they did not suspect a deliberate deception, one map was carried by a patrol of two armored cars that ventured into "no man's land" between the British and German positions. Inevitably, they attracted German artillery fire. In a carefully rehearsed scenario, one car "broke down" with shells exploding all around. The crew abandoned their vehicle and were carried to safety by the second armored car.

In the abandoned armored car, the Germans found a battered copy of the map, covered in sand and marked with oil stains. A second copy was left in the wreckage of a jeep that had been blown up by a German mine in another part of "no man's land." A third was among documents in an officer's briefcase, which was left on a seat in a Cairo bar known to be

With copies of the maps appearing from so many different sources, there was every reason to hope that the Germans would accept them as genuine, issue copies to their troops, and plan their operations accordingly. At the beginning of September 1942, the German attack began, and within hours it became clear that the enemy was following the British disinformation plan to the letter. The false maps led the Afrika Korps through concealed minefields and into the soft sand in front of the Alam Halfa ridge, where their stranded tanks were picked off by British tanks and anti-tank guns, firing from concealed positions along the crest line.

After three days of mounting losses, the Germans were forced to withdraw. This was Rommel's last attempt to reach Cairo, and from then on the balance would tilt against him. Soon it would be the turn of the British to attack at Alamein. Both sides knew the onslaught must be delivered soon. For the German defenders, the only questions would be where and when?

frequented by German agents. When military police entered the bar to search for the case, they found that its contents were missing. Officers were told that a certain Major Smith had been court-martialled for negligence in losing confidential papers, and bazaar gossip hinted that he had been executed by firing squad, information which the German agents were likely to overhear.

BELOW A tank is camouflaged as a supply truck to help conceal Allied strength and intentions.

TOP Lifeblood of desert fighting: British troops refill cans at a water supply point.

In the long and tortuous history of campaigns of misinformation and deception, one stands unchallenged as the most complex, ambitious, and successful of all time. This is the Fortitude deception campaign mounted by the British and Americans in World War II, which actually resulted in a whole German army being kept out of the Normandy fighting until it was too late. To do this, the Allies had to use double agents, electronic intelligence, radar intelligence, communications intelligence, and false intelligence in a cleverly orchestrated campaign.

The dilemma facing both the Allies and the Germans was deciding where the cross-channel invasion would land on the coast of France. There were only two sensible options: the Pas de Calais and Normandy. The former offered the shortest sea crossing from Britain, and the shortest route to the German heartland, but contained the heaviest and most formidable units in the German order of battle. The latter, while being less heavily garrisoned, required a longer and more dangerous sea crossing, and meant that much greater distances had to be covered before Germany itself could be threatened.

In fact, the strength of the Pas de Calais defenses had led the Allies to decide on Normandy from the outset. However, a massive deception campaign was mounted to assure the Germans that the landings would take place in the Pas de Calais. First of all, the network of German agents operating under British control sent a series of carefully coordinated messages that enabled the Germans to build a picture of massive forces being assembled in the south-east of England, ready for landings in the Calais area. At the same time, no signs were revealed of the real buildup of troops in the south-west, in preparation for Normandy.

The spies' messages were reinforced by all other channels of information. Large numbers of false radio signals were sent from transmitters in the south-east, and genuine messages from Montgomery's headquarters in Hampshire were sent by landline to a transmitter in Kent, so that these too could be added to the evidence for the buildup in the south-east. In case reconnaissance aircraft were sent over from France, dummy tanks, guns, aircraft, and landing craft were massed in the fields and river estuaries in the east. When Allied aircraft began the huge campaign of pre-invasion attacks on German forces in

ABOVE Maintaining the Fortitude deceptions prior to D-Day meant providing huge numbers of dummy forces in south-eastern England to attract the attention of German reconnaissance flights: a dummy Spitfire fighter **(TOP)**, a sketchy framework for a dummy Hurricane fighter-bomber **(ABOVE CENTER)**, and a framework that could be fitted to a Jeep to make it look like a Sherman tank from the air **(ABOVE)**.

France, the planners took care that three sorties were made against the Pas de Calais for every two against targets in Normandy.

The heart of the deception was to be the entirely fictitious FUSAG, or First US Army Group, the force earmarked for the Pas de Calais landings, to be commanded by the very real General George S. Patton. Calculated leaks revealed his presence in eastern England, and details of individual American units were fed back by the German spies, directed by their Allied spymasters. Even the most trivial snippets of information added to the overall impression. In neutral Geneva, British agents bought all available copies of the Michelin maps of the Calais area, an action carefully noted and reported by their German opposite numbers.

The picture was reinforced by a genuine German general too—Hans Cremer, former commander of the Afrika Korps, who was being repatriated to Germany in May 1944. En route to London from a prison camp in Wales, his car was diverted through south-west England. Because all location names and signposts had been removed as an anti-invasion measure in 1940, there was nothing to tell him that he was not being driven through southern and south-eastern England, and he was able to see signs of the huge buildup of men, weapons and materials ready for the invasion.

When the landing forces actually sailed, a new deception campaign swung into action. The German radar stations watching the entire Channel coast were destroyed by Allied fighter-bombers, except for two in the Pas de Calais area. These were deceived by two precision bomber squadrons of the Royal Air Force, which spent the entire night flying an exact pattern that allowed them to drop metallic foil strips to create echoes on German radar simulating a huge invasion force approaching across the Channel at low speed. In case the Germans sent naval craft to investigate this ominous radar picture, it was reinforced by a small fleet of naval motorboats, each towing a large radar-

LEFT General George S. Patton, commander of the entirely fictitious First US Army Group in the D-Day deceptions, and later the architect of the Allied armored breakout from the Normandy beachhead.

OPPOSITE Dropping "Window" (metal foil strips) to jam German radar during an RAF Bomber Command raid on Munster, September 12, 1944.

reflecting balloon. Sound equipment aboard the boats reproduced the noise of engines, anchors being dropped, loudspeaker commands, and bugle calls.

On land, the deception was intensified when large numbers of dummy parachutists were dropped all over the invasion area, loaded with firecrackers to simulate small-arms fire. Small Special Forces teams were also dropped with record players and loudspeakers to create the sounds of pitched battle in the darkness, confusing the German defenders as to the positions of their enemies.

It could be thought that the value of all these deceptions would have vanished once the landings had taken place. At that stage, under normal circumstances, the Germans would have known the truth, and could have taken action accordingly. But it was here that the most brilliant twist in the whole campaign was mounted. Messages from their agents in Britain were used to remind the German spymasters of the huge forces still remaining in south-eastern England and not used for the Normandy landings. Their worries were compounded by vital information from their most respected agent of all, Garbo (see Chapter 6). Two days after the invasion, Garbo had been ordered by his Allied controllers to report to the Germans that two units, the British Third Infantry Division and the Guards Armoured Division, were on their way to Normandy, on the grounds that the Germans would soon discover the units' presence for themselves, and this information would heighten the agent's standing and credibility.

At that moment, when they were on the point of ordering seven divisions to move from the Pas de Calais to crush the Normandy invasion, Garbo sent another message, a marathon two-hour signal, which explained that Normandy was intended as a feint attack, intended to lure the powerful German Fifteenth Army from its positions in the Pas de Calais. Once the Germans reacted to the bait, and moved their most powerful forces westward, the real Allied blow would fall behind them, and with it the Pas de Calais, and the direct route to Germany.

This inspiration turned the tables completely. The more determinedly the Allies attacked in Normandy, the more it would appear that they were trying to provoke the Germans into reacting. Only by holding back their most powerful units could they frustrate the Allied plan, so they kept the Fifteenth Army where it was for six vital weeks. By the time the Germans finally realized that FUSAG was a complete fabrication, Patton was in Normandy, leading the breakout from the bridgehead that eventually would sweep through northern France and help drive the Germans back to the borders of the Reich.

The Alamein disinformation campaign

General Montgomery's Eighth Army had two basic options—to attack the northern end of the German front or the southern. With a limited number of tanks still at his disposal, and with a chronic shortage of fuel, it was vital for Rommel to predict his opponents' intentions. Whenever the Luftwaffe could mount reconnaissance patrols in the face of growing Allied air superiority, they would be looking for telltale signs of attack preparations on the flat and featureless desert, which made concealment all but impossible.

All the indications noted by the German airmen added up to one clear message—the attack would come at the southern end of the front. First of all, the largest supply dumps were being assembled in the south, with massive numbers of tanks and guns, ready for the coming offensive. This was confirmed by high levels of Allied radio traffic, which direction finders revealed were being transmitted by units in the southern sector. Most reassuring of all was evidence that the British were building a water pipeline across the desert to supply these southern forces. By measuring its rate of progress, the Germans knew that the attack could not be delivered until early November.

Unfortunately, the picture was entirely false. Most of the British tanks and guns were really in the north, camouflaged as trucks and old supply dumps. The tanks and guns in the south were dummies, made from wood and canvas, and by disguising trucks. Radio transmissions between units in the south were mostly false, while genuine units in the north maintained radio silence as far as possible. Even the water pipeline was a dummy, built from flattened water cans, but looking genuine enough from an aircraft flying overhead.

The result was that when Montgomery's troops attacked the northern end of the front on October 23, 1942, Rommel's armored regiments were in the wrong place. So was Rommel; he was receiving hospital treatment in Germany. For the vital phase of the battle, the Afrika Korps was without his inspired and quick-thinking leadership, and after days of bitter fighting, the Germans began to pull out of the Alamein position. Their retreat would take them the length of the North African coast to distant Tunisia, and to surrender and capture seven months later.

The Alamein deception campaign, and the even more ambitious Fortitude deceptions, which protected the Normandy landings in 1944, laid the foundations for later

BELOW One aspect of the main Alamein deception campaign was the construction of a dummy water pipeline to mislead the Germans over the place and time of the British attack.

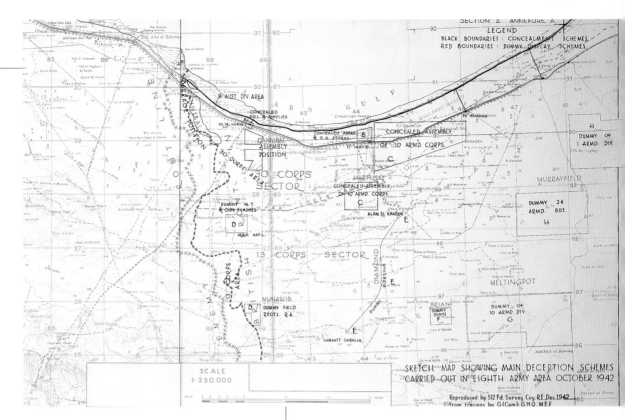

SKETCH MAP SHOWING MAIN DECEPTION SCHEMES
CARRIED OUT IN EIGHTH ARMY AREA OCTOBER 1942

SCALE
1:250,000

Reproduced by 512 Fd. Survey Coy. RE Dec.1942
from tracings by G (Cam) G.H.Q. M.E.F.

SECTION 2. ANNEXURE A
LEGEND
BLACK BOUNDARIES : CONCEALMENT SCHEMES.
RED BOUNDARIES : DUMMY DISPLAY SCHEMES.

ABOVE Papers from the headquarters of the Middle East Forces show details of the main deception schemes devised for the Alamein campaign in the fall of 1942.

LEFT A dummy pumphouse and reservoir were part of the water pipeline project, being designed to look completely convincing from the air.

BELOW General Bernard L. Montgomery, the new commander of the British Eighth Army in North Africa, and victor of the Battle of Alamein over his German adversary, Rommel.

misinformation campaigns that have continued to the present day. Compared with the inspired actions of individual commanders like Washington and Marlborough, today's false leads have to present a much more complex and multifaceted picture, through all the many different channels of information available to an opponent. Spies, radio traffic, reconnaissance flights, satellites and newspaper reports all have to tell the same, ultimately believable, story. If several independent sources confirm that a particular piece of information is true, it is likely to be accepted as genuine.

Operation Mincemeat

Once the Germans had surrendered in North Africa, the need for a new disinformation campaign emerged. Although the Allies had decided to invade the southern coast of Sicily to provide a base for later landings on the Italian mainland, it remained essential to keep the Germans guessing. This meant providing them with false information, but in such a way that its veracity would not be called into question.

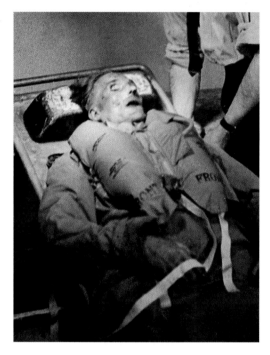

The intention was to convince the Germans that the Allies were planning to invade Greece, as a plausible alternative to Sicily. The problem was that a careless radio message would not be acceptable, and if the Germans saw through the deception, they might conclude that Sicily was the true target. They would know only too well that really important messages were carried in locked briefcases, chained to the wrists of couriers. So how could a courier be made to fall convincingly into enemy hands, with the vital information still intact?

Eventually, the Allies decided that the only realistic scenario would be to suggest to the Germans that the aircraft carrying a courier and his secret documents had crashed into the sea. This would account for a case containing sensitive documents being washed up on the coast of Spain, where Allied aircraft regularly flew offshore and where there was a strong German intelligence presence, who were likely to be told of the documents' discovery. One snag remained, however—the courier's body would have to be washed ashore still chained to the case.

This meant finding a recently dead body, in a condition that would suggest death by drowning, following an airplane crashing into the sea. Enquiries to London coroners produced the corpse of a young man who had died from pneumonia, which would produce similar physical signs to drowning. This allowed the deception plan, known as Operation Mincemeat, to be set in motion. The body was clothed in Royal Marines uniform and given documents that identified him as Major William Martin. In his pockets were letters from his father, girlfriend, and bank manager, together with ticket stubs from a recent performance at a London theatre. His briefcase contained letters from the office of the Chief of the Imperial General Staff and from Lord Louis Mountbatten to officers on the staff of General Alexander, commanding in the Mediterranean.

TOP The body of "Major William Martin" of the Royal Marines, dressed in uniform with a life preserver and showing all the post-mortem signs associated with drowning after an airplane crash.

ABOVE Some of the carefully selected evidence carried on the body to establish "Major Martin's" identity.

The content of the letters was vital. Rather than simply announcing that the Allies were intending to land in Greece, they discussed detail changes in the plan, of which the recipients would already be aware, but which would come as a revelation to the Germans. They cunningly referred to the plan under the genuine covername for the Sicily landings, Operation Husky.

The theory was that any references German agents might pick up to this operation would add to their conviction that the information was genuine. In an additional master stroke, they also referred to a deception plan that was being mounted to convince the Germans and Italians that the landings were to be made on Sicily instead, and to certain operations the Allies could not hope to hide, like the practice landings on the Tunisian coast and the heavy attacks on Sicilian airfields, as being designed to add credibility to this plan.

The body of "Major Martin" and the vital briefcase were loaded onto a Royal Navy submarine, which surfaced off the Spanish coast on the night of April 30, 1943. The corpse was lowered into the water at a spot where the currents would carry it ashore near the port of Huelva, where German diplomats were particularly active.

After two weeks of progressively stronger British representations for the return of the body, and the case and its contents, the Spanish authorities handed them over. All the documents were still in their sealed envelopes. However, laboratory examination showed that they had been opened and resealed, implying that copies of the papers were already on their way to Berlin.

By the time the Sicilian landings went ahead, 10 weeks after the submarine's cargo was washed up on a Spanish beach, there were only two German divisions defending the coast, reinforced by an Italian garrison of doubtful morale. Powerful German armored divisions had been switched to Greece and Crete, together with naval units that might have disrupted the landing forces. Rommel set up his Greek headquarters on the very day that Allied troops were wading ashore on Sicily, and although the campaign to liberate the island was hard-fought, it ended with the Germans having to flee to the mainland.

After the war, captured documents showed that the Abwehr (German Military Intelligence) had been doubtful about the authenticity of the documents, but Adolf Hitler had been convinced that they were genuine, and the Operation Mincemeat deception campaign was proved to have been totally successful.

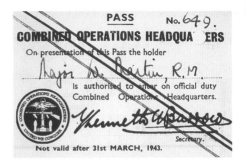

ABOVE "Major Martin's" pass, admitting him to Combined Operations Headquarters, accounted for the importance of the documents he was carrying and helped establish that they were genuine.

BELOW The "Martin" deception helped to mislead the Germans and caused them to divert forces that otherwise could have opposed the Anglo-American landings on Sicily in 1943.

The Twenty Committee

The British counterintelligence organization achieved an astonishing success during World War II, capturing every one of the German agents sent to the UK to relay much-needed information back to their masters. All too often, arrests were made because German agents had been badly briefed, spoke poor English, or were ignorant of wartime routines or restrictions. Others had incorrect or inadequate papers.

One of the most difficult hurdles faced by any agent landing by parachute or submarine was that they would find themselves on an island where strangers were immediately suspect, and where foreign nationals not wearing uniform were much rarer than in Germany, where the national economy depended on huge numbers of workers from the occupied territories. In addition, the Germans also employed agents who intended to work for the Allies from the very beginning, and who actually sought out the authorities as soon as they arrived in Britain, but their loyalty was never doubted by their original employers.

Once the first agents had been captured and turned, the British were able to use them to pass on information they wanted the Germans to believe. The British XX (for "Double Cross," but known as "Twenty" in reference to the Roman numeral) Committee, which capitalized on the 100-percent success record in capturing German agents, was able to monitor the growing spy network that the Germans believed they had created in Britain. This enabled them to avoid several of the pitfalls that might otherwise have caused mistakes to be made.

For example, some agents were ordered to carry out active sabotage operations, like the detonation of a bomb in the de Havilland aircraft factory at Hatfield in Hertfordshire. Other agents, working on British instructions, were able to report that these attacks had been successful, or that in some cases the agents responsible had been killed or captured. Bomb damage was carefully simulated, so that German reconnaissance planes could confirm the success of the operation.

On one occasion, in November 1941, the XX Committee sent agents to set off an explosion in a food store at Wealdstone in north-west London. Two elderly security guards had to be lured to one end of the premises, while a local policeman not party to the plot had to be evaded, before the bomb could be set off. Eventually, firefighters were called, and reports in the

TOP One of the most successful double agents, former safebreaker Eddie Chapman, was in prison on the British island of Jersey when it was captured by the Germans in 1940. The Germans recruited him and sent him to Britain as a spy.

ABOVE Chapman's notes on the airfield at Hatfield, north of London,

newspapers reached the Germans, which reassured them of the value of their agents.

Cover stories were carefully adjusted to limit the flow of information. One German agent, codenamed Tate, was sent so much money that he could have moved all over Britain to gather information from sensitive locations. His British controllers told him to report that he had been questioned by the police for failing to register for military service, but that he had found work on a farm to gain exemption from the army. This would restrict his free time to weekends, and his travel to a much smaller area.

Some German agents refused to cooperate and were imprisoned or executed. In some cases, their actual fate was revealed in messages sent back to Germany by the double agents. In others, where their background was particularly valuable, British radio operators sent and received messages on their behalf, so the Germans never suspected that they had been caught. One agent, an English safebreaker called Eddie Chapman, codenamed Fritzchen by the Germans, was actually recalled by them to report in person. He had been in prison in Jersey when the Germans occupied the island and had volunteered to work for them as an agent in England.

The British gave Chapman the codename Zigzag, and it was decided to let him make the journey to reinforce the credibility of the whole network of agents. He was given work as a steward on a British merchant ship bound for Lisbon, where he deserted and reported to the Abwehr. He was praised for the value of his work, sent to the Abwehr office in Oslo to advise on sabotage methods, and was parachuted back into England a second time in June 1944, with orders for another sabotage mission. Once again, he reported to the authorities and was able to give them valuable, up-to-date information on wartime conditions inside Germany.

ABOVE When Chapman arrived in Britain, he gave himself up and volunteered to work for the British as a double agent, under the codename "Zigzag."

BELOW "Zigzag's" diagram and notes on the layout and facilities of the airfield and aircraft factory at Hatfield in Hertfordshire, where Mosquito fighters and bombers were built and tested.

BACKGROUND Pages from "Zigzag's" notebooks on the details of the Hatfield airfield.

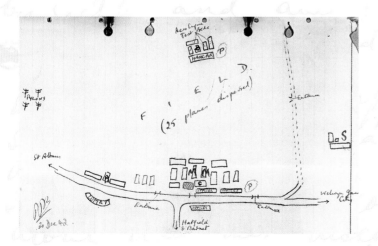

Operation Nordpol

Once a spy is operating on alien soil, he or she is beyond the range of support from the organization responsible for the mission. The threat of interrogation, torture or death following capture is a powerful inducement to operate on behalf of the other side. However, if the spymasters know that their agent has been turned, the danger can be contained and even used to advantage.

To provide a warning that an agent had been captured, during World War II, the British Special Operations Executive gave their radio operators pre-arranged errors to incorporate in messages, which would confirm that they were still operating freely. A completely correct message would show that they had been captured and were sending information under German direction.

Yet in the greatest wartime disaster suffered by SOE, these checks apparently failed. Hubert Lauwers, an SOE radio operator transmitting from the Hague, was captured by the German Abwehr counterintelligence organization, headed in the Netherlands by a former tobacco merchant, Major Hermann Giskes. Giskes ordered Lauwers to continue transmitting signals to London in a payback operation that he christened Operation Nordpol (North Pole). Lauwers was ordered to transmit his messages on schedule, and did so on the assumption that SOE would at least know what had happened to him, because he sent the messages with no mistakes.

Incredibly, the mistakes went unnoticed or were ignored in London. As a result of Lauwers's signals, arrangements were made for drops of arms, supplies, and other agents to join the Dutch networks. The Germans picked up the supplies and arrested the agents. For the best part of a year, the British delivered thousands of weapons, tons of explosives, and more than 50 agents straight into enemy hands. Although all of Lauwers's messages were sent back without the deliberate mistakes that would have proved their authenticity, the Dutch Section of SOE took no action. The desperate agent tried, at great personal risk, to include the word "caught" split into two fragments at the beginning and end of each message, but even then London took no notice.

Only when two of the captured agents, Peter Dourlein and Ben Ubbink, managed to break out from their German prison and cross Occupied France to the safety of neutral Spain, before reporting to London, did Operation Nordpol come to an end. Even then, Giskes used one of the captured transmitters to send a message to London saying that the two men had been revealed to be the traitors, which had led to all the agents and supplies being captured. As a result, when Dourlein and Ubbink reached London, they were arrested and thrown in jail.

SOE's Dutch operation never recovered from this disaster. Some postwar rumors suggested that the whole affair had been a deliberate plan to induce the Germans to believe false information, sent with the captured agents, but an inquiry revealed that, in fact, it was due to plain incompetence. The Dutch Section had committed a series of errors, like dressing different agents in identical clothes bought in London shops, rather than in genuine Dutch clothing. They had supplied agents with silver coins no longer in circulation, and with poorly forged papers that were discovered at the first security check. They had told their agents that the radio transmitters were undetectable by German direction finders, which definitely was not the case. Finally, they had not told the agents that the transmitters would only work with large outdoor antennas, which were very difficult to conceal. In the light of this catalog of errors, the Nordpol disaster fitted the pattern all too well.

BACKGROUND Miniaturized and easily concealed by the standards of the 1940s—the transmitter/receiver issued to German agents by the Abwehr.

The "black" radio campaigns

Perhaps the ultimate in false information is a completely bogus broadcasting service, which disguises the country sponsoring its transmissions. This is the realm of so-called "black" radio, operated by both sides in World War II to considerable effect. The Germans set up a station that announced itself as the New British Broadcasting Service. In sharp contrast to the anti-Allied propaganda of official "white" German radio, delivered by a team led by William Joyce, known at the time as "Lord Haw-Haw" because of his drawling tones, the station pretended to be on the side of the British people, but against their government—and definitely against the Germans.

The theory behind "black" propaganda of any kind is that people will pay more attention to material that ostensibly is on their side. Therefore, any subtle messages it seeks to put over are more likely to be accepted at face value and can be used to damage morale. However, this can only occur when two requirements are satisfied. The material—written, spoken or visual—must be entirely convincing, and it must have a background of truth to boost its credibility.

The New British Broadcasting Service failed the first requirement. It always sounded oddly unconvincing to British listeners, who failed to pay it much attention. German "black" radio broadcasts to France, as the country collapsed under the onslaught of the Panzer divisions in the summer of 1940, were much more effective, playing a crucial role in creating panic and maintaining deep suspicion of those genuinely in authority.

But the past masters at "black" radio broadcasting proved to be the British. Their first station was called Gustav Siegfried Eins, from German signallers' jargon for GS1, which it was thought would appear to stand for Secret Transmitter 1 or General Staff 1. It purported to be run by a group of dissidents who were entirely pro-German, and even pro-Hitler, but who were deeply opposed to the corruption and incompetence of the Nazi Party. Although the signals were transmitted from England, the station claimed to be operating from secret locations within the German Reich.

How was GS1 able to sound convincing? Because its key broadcasters were all native-born Germans, many of whom were anti-Nazi former prisoners of war who knew the slang and

TOP William and Margaret Joyce. William is wearing the uniform of a civilian in the Nazi party.

ABOVE A citation for the award of the German War Service Medal made out to Margaret Joyce.

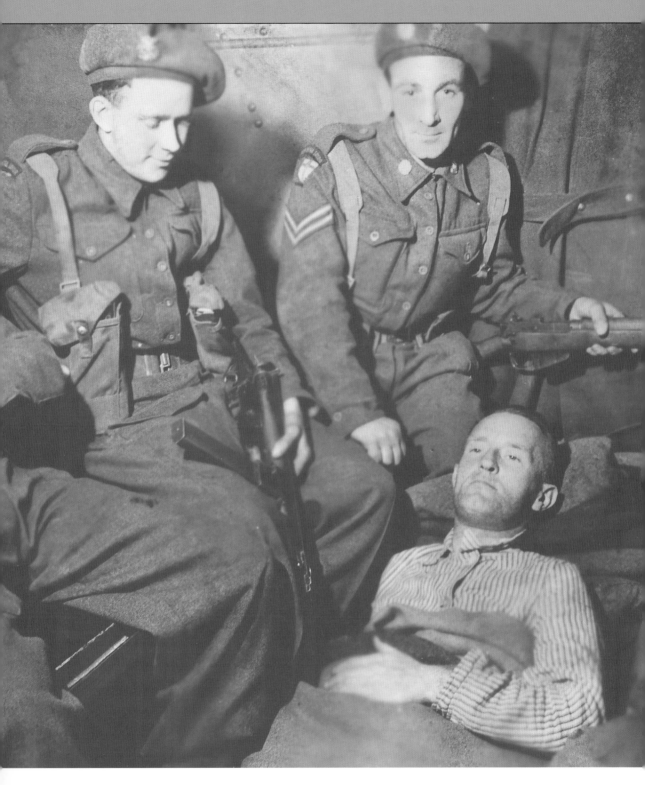

ABOVE William Joyce, popularly known as "Lord Haw-Haw" because of his drawling delivery when making German propaganda broadcasts to Britain during World War II, under arrest in May 1945.

routines of the German armed forces. Its news items came direct from the German Propaganda Ministry, through intercepted radio broadcasts to German newspapers. It rebroadcast speeches by Hitler and other leaders, which were recorded by the BBC's monitoring services from actual German broadcasts. But the key to the station's effectiveness was the subtle spin it put on quite genuine news items.

For example, a news release explained that the German authorities were worried that too many citizens would cash in their clothing coupons with the onset of winter, resulting in shortages. Gustav Siegfried Eins broadcast this item as a scandal story, explaining how the wives of top Nazi officials were being tipped off to buy their new winter coats to keep ahead of the shortages of warm clothing that would be suffered by the army on the Russian Front. Another news item referred to children being evacuated to special camps away from the bombed German cities. Gustav Siegfried Eins congratulated the doctors at one of the camps for managing to reduce the death rate from an outbreak of diphtheria to a level of just 60 children a week.

It was cruel, but it was effective, to the point where the Americans, who were not in on the secret at the time, recorded the broadcasts as evidence of dissident groups operating inside Nazi Germany. The British went on to set up two spurious German services' broadcasting stations—Atlantiksender (for the U-boat service) and Soldatensender Calais (for the army). Both were accepted as genuine by their listeners, which enabled the same carefully doctored news items, on a solid foundation of truth, to chip away at morale.

Even when the stations were officially revealed to be run by the Allies, they continued to damage morale, but in a different way. Fed with intelligence information from agents throughout the occupied countries, they could afford to be accurate over the most trivial of details. For a U-boat crew, setting off into the perils of the Atlantic battle, where survival depended on remaining undetected, hearing the latest football scores from their own base, coupled with a record request for a named member of their crew, to wish them good luck, was potentially devastating to their hopes of returning, and crippling to their operational efficiency.

ABOVE William Joyce in "disguise" as a Hitler look-alike.

BELOW A page from Margaret Joyce's German diary showing entries for the first weeks after the German surrender. On May 28 she wrote, "Quarrelled with Will and let him go out alone and he was arrested and shot at. I was arrested with all the (?) but they let them go. Some of the officers came and jeered at me but some were nicer."

147—150	MAI	1945

Sonntag *Nick's birthday*
27 ☿
Trinitatis *Attacked word in the afternoon*

•

Normalzeit: SA. 3.49 SU. 20. 6 — MA. 20.30 MU. 4.18

Montag *Quarrelled with Will & let him go out*
28 *alone & he was arrested & shot at.*
I was arrested with all the Begumgums
but they let them go. Some of the officers
came & jeered at me but some were nicer.
1940 Kapitulation der belgischen Armee

Dienstag

Falklands War deceptions

Many of the classic disinformation and deception campaigns belonged to World War II, but the art still remains an essential resource for military commanders seeking to influence conditions in their favor. In the 1982 Falklands War, the Royal Air Force went to considerable lengths to make the airfield at Port Stanley unusable to the occupying Argentine forces. By sending up relays of refuelling tanker aircraft, they made it possible for a lone Vulcan bomber to reach the Falklands after a marathon non-stop flight from Britain. It flew in over the islands and dropped a stick of high-explosive bombs across the main runway. Because this was done at night, there was no way of knowing how successful the raid had been until a high-level photo-reconnaissance mission could be flown on the following morning.

The result was very encouraging. As expected, there was a row of bomb craters right across the airfield, with one right in the center of the main runway, preventing its use by military supply aircraft. Therefore, there was no need to repeat the whole costly and complex operation on successive nights. Unfortunately, this was a clever Argentine deception. After the raid, and before first light, the occupying forces had sent earthmoving equipment out onto the airfield to repair the only crater that had affected the runway, but they created mounds of

ABOVE Royal Air Force delta-wing Vulcan bombers once carried Britain's nuclear deterrent, but in 1982 one was used to deliver conventional bombs in an ultra-long-range raid intended to deny Port Stanley airfield to the invading Argentine forces.

BELOW An aerial view of Port Stanley airfield, showing bomb damage on the runway. British supply drops were made from Ascension Island, a 9,000-mile round trip, until the airfield had been declared safe.

soil in a pattern that
would look exactly like a bomb
crater to a reconnaissance aircraft.

Ironically, this was a repeat of a classic
deception first used by the British in the desert fortress of
Tobruk during World War II. Because the survival of the troops
being besieged by Rommel's Afrika Korps depended on the
water supplies produced by the town's distillation plant, the
German Luftwaffe tried to bomb it repeatedly. After one
particularly heavy raid narrowly failed to inflict fatal damage,
the defenders dug fake bomb craters all over the site, darkening
them with a mixture of oil and coal dust to make them look
deeper. They also blew up an old cooling tower that was no
longer needed, painted the side of the main building to look as
if bombs had exploded inside, and scattered pieces of wreckage

BELOW A censored Argentine
military photograph shows earth
scattered on the runway by bulldozers
to suggest heavy damage.

all over the site. After German reconnaissance aircraft flew over to check the damage, the raids ceased, although the plant continued to produce drinking water for the duration of the siege.

At the very start of the Falklands War, the British had been able to mount a stunningly simple deception of their own, which effectively kept the bulk of the Argentine Navy out of the fighting. Before the Royal Navy task force reached the islands, the only British naval unit in the area was the Antarctic patrol ship *HMS Endurance*, which was equipped with a small Wasp helicopter.

While the unarmed *Endurance* was hiding from the Argentine landing forces amid the icebergs of the South Atlantic, her helicopter pilot flew the Wasp at wavetop height and held a radio conversation with the ship using the code letters for the Royal Navy nuclear-powered hunter-killer submarine *HMS Superb*, asking for information on sea and weather conditions, and arranging a rendezvous. At the time, *Superb* was 8,000 miles away in the dockyard at Plymouth, but the Argentines identified the code letters and the approximate position of the transmission, and their surface ships kept well away from what they perceived as the deadliest of threats.

ABOVE A soldier from the British task force examines the wreckage of an Argentine Air Force Pucara ground-attack aircraft, on the runway at Port Stanley airfield after the surrender.

BACKGROUND Argentine aircraft and bomb craters on the runway at Port Stanley in the Falkland Islands.

The temporary defector

One of the most valuable defectors ever for the United States, was Vitaly Yurchenko, a KGB officer who telephoned the Rome office of the CIA, located in the US Embassy on Via Veneto, on August 1, 1985, to arrange a meeting. When the meeting took place, he revealed his intention to defect, and made it clear that following a 25-year career with the KGB after transfer from the Soviet Navy in 1960, he had much valuable information to deliver. In particular, he revealed that a former employee of the National Security Agency, known to the Russians as Mr Long, had passed on details of a secret US Navy/NSA project named "Ivy Bells," which required a submarine to fit a tapping device to the Soviet military communications cable running across the bed of the Sea of Okhotsk. Checks revealed that the true identity of "Mr Long" was Ronald Pelton, who was arrested in June 1986 and who confessed to spying for the Soviet Union.

Yurchenko was also able to set the record straight on a suspected spy who, in fact, had been entirely innocent. Leslie James Bennett, who had worked for British signals intelligence during World War II, had emigrated to Canada in 1954, and had joined the Royal Canadian Mounted Police, eventually becoming head of the RCMP Security Service counterintelligence section monitoring Russian espionage operations in Canada. When a series of counterintelligence campaigns went wrong, Bennett came under suspicion as a newcomer. Although no direct proof was found, he was forced to resign, and later he emigrated to Australia. Yurchenko was able to reveal that the real author of these failures was a genuine Soviet agent within the RCMP Security Service named Gilles G. Brunet, a native Canadian whose father had been the first head of the Security Service!

Astonishingly, within a matter of months, Yurchenko was back in Russia. Following his defection, he had become increasingly unhappy about the restrictions placed on his movements in Washington, and about the possible fate of his wife and child. On November 2, 1985, he evaded his CIA minder by leaving a restaurant without warning, and rushed to the Soviet embassy. There he claimed to have been kidnapped and drugged by CIA agents, and four days after his escape, he was flown back to Moscow. At the time, it was considered possible that his defection had been a KGB ploy, but given the information he provided, the CIA concluded that this was a genuine case of a real defector having changed his mind and deciding to return home to face the consequences.

TOP Lawyer Fred Bennet (on courthouse steps) announces that his client, former National Security Agency communications specialist Ronald Pelton, is to appeal against his conviction on four counts of having sold information to the Russians on US intelligence gathering operations.

ABOVE Former Russian KGB officer Vitaly Yurchenko announces his intention at a Soviet Embassy reception to return to Russia.

Airborne Intelligence

ABOVE A scene from an 1837 painting by Jean-Baptiste Mauzaisse, showing the Battle of Fleurus in 1794, during the French Revolutionary Wars.

OPPOSITE TOP A World War I French observation balloon spotting for the artillery at Madenay on the Marne, July 6, 1917.

OPPOSITE BOTTOM Professor Thaddeus Lowe's observation balloon "Eagle" in the American Civil War, seen during a storm, with infantry, artillery, and supply wagons in the background.

From man's earliest experiments with tethered balloons, airborne observation has added another dimension to intelligence gathering, allowing military commanders their first real opportunity to answer the classic question of what was really going on over the hill. At the battle of Fleurus on June 26, 1794, the French General Morlot was able to watch the movements of his Austrian adversaries from a hydrogen balloon for 10 hours, sending his written orders down the balloon cable to officers waiting on the ground. Both sides in the American Civil War used observation balloons to report the movements of enemy troops. The Union in particular sent a whole series of balloon units into the field, including two that operated from the decks of a tug and an armed transport in the James River, thus forming the first carrier task force in history.

By the outbreak of World War I, both sides were making use of balloons to spot the fall of shot for artillery, allowing the gunners' aim to be corrected. However, the value of powered aircraft, in terms of the area they could cover on a single mission, soon became realized. At first, they carried trained observers who could report what they had seen with their own eyes, when they returned to base. Later, airborne cameras would photograph a wide area in very fine detail. The films were processed once the aircraft had returned, then were analyzed to produce the maximum amount of useful information.

Gradually the science of photo-interpretation was developed to make the best possible use of the material that aerial reconnaissance provided. In a world where trench warfare dominated the battlefield, details of enemy fortifications were revealed with absolute clarity. Fuel and ammunition dumps would indicate the preparations that had to be made before launching an attack; large groups of marching men could be seen in photos taken from a height of several thousand feet; and in one case, German aircraft took photos of tapes marking out the lines of German trenches, but laid behind the British lines for troops practicing the attack on the Somme fortifications in July 1916.

To hide the truth— camouflage

To combat the eye in the sky, a parallel science of camouflage was developed to hide sensitive details, or to mislead the enemy. Gun emplacements were concealed by foliage or netting, buildings were disguised with camouflage paint, and men could hide in trenches and underground bunkers. More difficult to conceal were the tracks left by vehicles on grass or soil surfaces, and those made by regular visits to concealed positions

by reinforcements or ration parties. Furthermore, both sides soon realized that, while a single photographic mission over a given area provided valuable intelligence, the wealth of information that could be gained from repeated visits was infinitely greater. Any changes in the picture could be used to discern the most carefully hidden troop movements and supply concentrations.

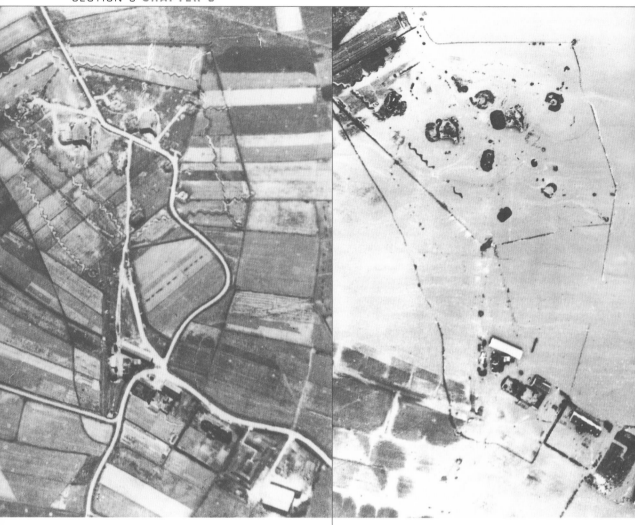

ABOVE German defensive positions near Westkapelle on the Dutch island of Walcheren, showing zigzag lines of trenches.

ABOVE After RAF Bomber Command had bombed the area on October 3, 1944, photo-reconnaissance pictures showed that the dikes holding back the sea had been breached, flooding the German defenses.

By World War II, a wide range of military reconnaissance aircraft had greatly increased the scope of airborne intelligence. Pictures of towns and cities could reveal industrial resources that would become attractive targets once the aerial war began. Photographs of airfields would reveal the strength and organization of defenses against which the bombing raids would have to be mounted. And when the raids had been completed, photo-reconnaissance missions would be sent to determine the extent and the effectiveness of the damage inflicted.

Whatever the undoubted value of photographic intelligence as a whole, there was one application that would change the shape of an entire campaign—RAF Bomber Command's night bombing offensive against the heart of Nazi Germany. Faced with reports from neutral observers that many heavy raids seemed to have missed their targets, in contrast to the enthusiastic reports from the bomber crews of widespread damage, the RAF fitted cameras to all their bombers. Each camera was triggered by a flash flare, which was fired when the bombs were dropped, revealing the exact whereabouts of the aircraft at the time.

The resulting information was not only very valuable, but also immensely discouraging. It showed that only a handful of bombers were landing blows anywhere near their targets, despite the undoubted courage and dedication of their crews. The almost insuperable difficulties of navigating all the way to Germany over a totally blacked-out Europe were preventing them from being certain of their position to within a few miles.

The photographs instilled a sense of urgency at Bomber Command. New navigational techniques were developed to improve the bombers' chances of finding their targets, using the most experienced crews to relay information on winds and courses to their less skilled companions. And in the longer term, radar-based navigation systems—like Gee, Oboe and H2S—were developed and fitted to bombers to guide them precisely to their targets in conditions of thick cloud and poor visibility.

World War II airborne intelligence gathering

During World War II, photo-reconnaissance developed far beyond the pioneering efforts made in the previous war. A new generation of high-performance aircraft, like the American P38 and P51, and the British Spitfire and Mosquito, were able to fly fast missions into the heart of Europe, take the necessary pictures, and return without falling prey to enemy fighters or anti-aircraft fire. Some missions were flown at high altitude, to cover a large area in a single pass, while others were at treetop level to give close-up detail of individual sites. Some low-level operations took place at dawn or dusk, when the sun was low in the sky, casting long shadows on the ground that revealed much detail about the true shapes of objects captured by the lens.

Multiple, power-driven cameras were able to shoot a series of exposures on a single high-

ABOVE An oblique high-altitude photograph, taken by an RAF reconnaissance aircraft, shows the French coastline and the estuary of the River Seine at Le Havre.

speed pass over a target, which enabled "stereo pairs" to be produced. By aligning adjacent prints taken on successive exposures and examining them through a stereoscopic viewer, the photo-interpreter could see a detailed three-dimensional image that revealed crucial facts like the heights of objects in the picture. The technique also allowed the shapes hidden beneath camouflage netting to be discerned, and could "see" through otherwise effective camouflage schemes that relied on two-dimensional images on the ground.

Later, a whole battery of systems was developed to strip away the protective layers of more ambitious camouflage schemes and reveal the truth that lay beneath them. Infra-red false color film, which reacted to the different levels of temperature in the scene exposed to the camera, could distinguish between dead and living foliage. Plants that had been cut and laid over a tank, truck or gun position stood out from the living landscape

ABOVE The Lockheed Lightning twin-boom fighter-bomber had the range to provide US forces with photo-reconnaissance coverage of a wide area of Western Europe.

with clarity. Grass and plants that had been crushed by vehicles showed up clearly, indicating where troops and their vehicles had been hidden. And living soldiers gave off the most obvious signals of all, compared with dummies intended to give a false position extra credibility.

Deceiving the photo-interpreters

On the other hand, it was possible to use the very effectiveness of photo-reconnaissance to mislead those who employed it. In the Allied deception and misinformation campaigns before the Battle of Alamein and the Normandy landings (see Chapter 7), a large number of dummy guns, vehicles, and airplanes had to be used to suggest a buildup of forces in a particular area. Yet if these had been genuine, they would quickly have been camouflaged to hide them from enemy aircraft. Because of this, it was possible to use the same dummies over and over again.

First, they would be left in the open on a particular site to suggest that they had only just arrived. Then they would be disassembled as if carefully hidden on the spot, with just a hint here and there of badly applied camouflage to leave a gun barrel partly exposed or signs of badly concealed digging. Once it was thought that the enemy had accepted that they had been covered up, the dummies could be placed on another site, and the whole process repeated, maximizing the value of the dummies far beyond their actual numbers.

In the postwar years, new reconnaissance systems would give air forces even greater coverage of the ground below, and allow new types of information to be gathered. Infrared line scan systems would show target areas with perfect clarity at night and in poor visibility, building up a picture that would show all kinds of detail. Aircraft with engines running on an airfield ramp could be clearly distinguished from dummies, or planes that were not running. Fuel storage tanks that were empty showed up differently from those that were full. Dispersal areas that had contained aircraft, which had recently taken off, were warmer than the surrounding concrete, and so on.

Radar coverage enabled huge areas to be covered in less precise detail. For the whereabouts of shipping, a crucial indicator in establishing trade patterns or preparations for hostilities, a high-flying jet could record a radar map of the whole Mediterranean in just two sorties. As in all these reconnaissance missions, regular coverage established patterns, and deviations from those patterns indicated a potentially dangerous situation.

While high-performance aircraft and more powerful reconnaissance techniques had tilted the balance in favor of the watchers over those whom they were watching on the ground, the situation was redressed with the development of more formidable anti-aircraft defenses in the shape of surface-to-air missile (SAM) systems. Even then, however, strenuous efforts were made to build reconnaissance planes that could evade the missile defenses.

ABOVE AND BACKGROUND Camouflage sheeting and other concealment measures are intended to break up the outlines of factories and similar vulnerable targets so that they fail to show up in reconnaissance photos.

Higher, faster, and stealthier—the rise of the spyplane

The first of the specialized spyplanes was the American Lockheed U2, developed in the 1950s. This was designed to fly at heights above 70,000 feet, to take it clear of the fighters and anti-aircraft missiles of the time, and it was made from the thinnest possible sections of metal to keep weight to the minimum. It had long thin wings for maximum lift, and its basic design configuration was similar to that of a powered glider. In fact, once it had used its turbojet engine to climb to its operational height, it would glide for long periods through the thin air of the stratosphere, ensuring maximum range and endurance, and minimum vulnerability to the defenses of the countries over which it flew.

The SS-X-24 and Flight KAL 007, USA, 1983

By 1983, American technology had developed to the point where it could provide a close-up view of Russian efforts to test fire each new generation of its missiles. That summer, the latest Soviet intercontinental ballistic missile was the SS-X-24, a formidable weapon fitted with multiple, independently targeted warheads. This was expected to be fired from the Plesetsk test area in north-western Russia, right across the Soviet Union to hit the Kamchatka Peninsula in the Soviet far east, on the last day of August, and a whole battery of American resources was in place to monitor the event.

A specially equipped American RC135 intelligence aircraft had taken off from a military airfield in Alaska, carrying sensors to record data from the Soviet missile as it re-entered the earth's atmosphere on its way to impact the target area, backed up by two different camera systems to photograph the dummy warheads and decoys ejected from the missile in the final phase of its flight, and to measure the size of the warheads as a guide to the weapon's potential destructive power. Other observations were being made by sophisticated radar and SIGINT equipment based in Alaska and carried aboard two US warships in international waters off Kamchatka. These were backed up by a series of listening posts in Alaska and Japan, which monitored all Soviet military communications channels to gather yet more information.

All was ready for this vital test of new Soviet weaponry, when an innocent passerby entered the scene, in the shape of Flight KAL007, a Korean Air Lines Boeing 747 flying from Anchorage in Alaska to Seoul in South Korea, with 240 passengers and a crew of 29. Tragically, at such a sensitive moment, a series of errors by the flight crew was causing the aircraft to veer progressively farther from its intended course. It was curving away to the north, crossing the extremely sensitive Kamchatka Peninsula, and was about to fly through some of the most heavily defended airspace in the world.

At first, the American communications monitors believed that the messages overheard from Soviet fighter pilots and ground controllers indicated that they were practicing their reaction to an American aircraft intruding into their airspace. In fact, they were in deadly earnest. The track of the Korean airliner had taken it across the peninsula, then out over the sea, whereupon the Soviets had relaxed. With the crew still unaware of the deadly peril they faced, the airliner continued, crossing the coast of the island of Sakhalin and putting the Russians back on to full alert.

The American SIGINT facility at the Misawa air base in Japan picked up signals from a Soviet radar station on Sakhalin, reacting to the approach of an intruder at 2:43 am local time. The voices of Russian fighter pilots could just be heard for a little over an hour, then all communications ceased at 3:47 am. To the listeners at Misawa, it was all a mystery. But to another American signals intelligence unit in Japan, the picture was grim. They had heard a Russian phrase signifying the launch of an air-to-air missile, followed by another confirming the target as destroyed. At about the same time, contact with KAL007 was lost. All 269 aboard the plane had died, killed by a deadly mistake in the international espionage battle.

In the end, the SS-X-24 was not launched that night, and no information ever leaked out from the Soviet Union to provide even the remotest justification for the loss of all those lives, which had resulted from a case of mistaken identity.

The U2 entered service in 1956, carrying a new generation of cameras developed by the man responsible for the Polaroid instant-print system. The cameras were special long-focus instruments, which were sighted through seven apertures in the U2's skin to produce extremely detailed, high-quality pictures of a 125-mile-wide strip of the ground more than 10 miles below. In one test shot, taken from 55,000 feet, the pictures included a golf course— successive enlargements made it possible to count the golf balls on one of the greens.

U2s flying from bases in Britain, Germany, Turkey, Pakistan, and Japan were able to cover huge areas of the Soviet Union and bring back extremely valuable pictures. One of their earliest successes was to correct the impression of Soviet airpower given by the Soviet Aviation Day in July 1955, which was marked by the traditional flypast of military aircraft over Moscow's Red Square. At a time when opinion was divided over whether manned bombers or missiles were the best way to deliver nuclear weapons, diplomatic and military observers at the event were appalled to see huge numbers of Russian bombers flying over in apparently endless formations. Yet the U2s' coverage of the Soviet aircraft industry showed beyond doubt that the huge force could not have been assembled using existing resources, and the impression must have been created by a much smaller formation of bombers, circling out of sight and making repeated passes over the capital.

In the end, the fateful mission of Francis Gary Powers (see Case Study No.36) revealed the vulnerability of the U2 to improved air defense systems. Although U2s could still deliver spectacularly valuable information in more restricted theaters of operation (see Case Study No.39), what was needed for the future was a series of new designs that could snatch back the initiative from the defenders.

One way of overcoming the defenses was to use speed. Lockheed's SR71 Blackbird, which was revealed in 1964, could more than match

ABOVE A Lockheed SR-71 Blackbird high-speed reconnaissance aircraft seen from a KC-135 tanker aircraft as it approaches the tanker's flight refueling boom. When the two aircraft are close enough, one of the tanker's crew can steer the boom to make contact with the SR-71's refueling point to replenish its tanks and extend its range.

the altitude performance of the U2, and could fly at up to three times the speed of sound. While long-distance overflights of the Soviet Union were ruled out on political grounds, these aircraft could bring back photographic and electronic information from other sensitive areas, relying on height and speed for protection against hostile missiles and aircraft. Later designs, which combined shapes and surface finish to minimize radar reflections, would employ stealth as an alternative means of evading enemy defenses. However, by the time these sophisticated aircraft became available, the global reconnaissance role had been taken over by satellites, operating at heights far above those covered by the boundaries of national airspace. Reconnaissance and intelligence gathering had truly reached a new dimension.

Hitler's revenge weapons

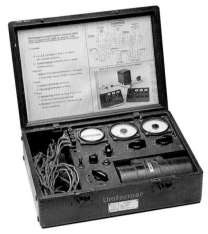

ABOVE A portable test kit for the German V2 rocket, used to carry out pre-launch checks before firing. The V2 was the world's first ballistic missile and could deliver a one-ton explosive warhead over a range of several hundred miles.

During World War II, Allied intelligence began to receive disturbing reports from observers and agents in Nazi Germany, which referred to the development of frightening new types of rocket weapon at a secret site on a remote island off the Baltic coast named Peenemünde. To provide confirmation of this research, RAF photo-reconnaissance Spitfires were sent to bring back visual evidence from the site. At first, it was clear that a lot of construction work had taken place for some highly technical activity, but it was not until photographs were produced by a sortie on June 12, 1943, when Professor R. V. Jones, head of Air Ministry Intelligence, spotted a silhouette of a large rocket lying on its side on a railroad flatcar, that it became clear that this was indeed the German rocket development center. Another sortie, on June 23, produced pictures with even clearer views of rockets on the site.

In fact, these rockets were prototype V2s, the ballistic missiles that would be used against London, Paris, and Antwerp in the closing year of the war, and which later would provide the first stages of the postwar American and Russian missile and space programs. In an attempt to nip the German rocket threat in the bud, a force of 597 RAF bombers was sent to bomb Peenemünde on the night of August 17/18, 1943. Although 41 aircraft were lost, the destruction severely delayed the rocket program.

As a direct result, the Germans switched the later test and development program for the V2 to a firing range at Blizna, in Occupied Poland. In time, this too was discovered by an ongoing photo-reconnaissance campaign, which covered the entire German railroad system in the area. It was almost impossible to conceal the long flatcars with their unmistakable cargoes, and repeated coverage enabled their progress to be traced from the Baltic coast to the test establishment in central Poland.

Nevertheless, other work continued in the neighborhood of Peenemünde, and a series of messages from agents, reinforced by clues in deciphered Enigma signals, revealed that the Germans were also hard at work developing a pilotless flying bomb, effectively the world's first cruise missile. Reconnaissance missions over the French coast had produced photos of structures that appeared to be catapults, and these were thought to be linked to the mysterious new weapon, which eventually was dubbed V1 by the Germans.

Although other evidence had been forthcoming, including reports from the units responsible for monitoring the range and accuracy of test launches over the Baltic, and radar signals that showed these missiles on site, they had never appeared in any photographs. Until a sortie by an RAF photo-reconnaissance Mosquito on November 28, 1943, over the village of Zempin, close to Peenemünde, where similar catapult-like installations had been discovered. When the prints from that mission where examined, there on one of the catapults was the small silhouette of the first V1 ever pictured.

Close to each catapult installation at Peenemünde and Zempin was a strange long building with a curved end, similar in shape to a ski lying on its side. Although their precise purpose was unknown, these buildings were clearly linked with the V1 launch catapults, and when reconnaissance photographs of the Channel coast of northern France showed a whole series of sites with these ski-shaped buildings, they were made high-priority targets for Allied bombers and fighter-bombers. The operation against them began with a massive strike by 1,300 bombers of the US Eighth Air Force, which dropped a total of 1,700 tons of bombs.

Nevertheless, the V1 campaign against London began a week after D-Day, and 8,617 flying bombs had been launched against the city by the time the advancing Allied forces pushed the Germans back out of range. These killed some 5,500 people and inflicted serious injuries on a further 16,000. However, without the dedicated work of the pilots and photo-interpreters in identifying the purpose and whereabouts of the "ski-sites," so that as many as possible could be destroyed before the campaign began, both totals would have been much higher.

BACKGROUND An aerial photo showing V2 rockets on test stands at the Peenemünde research establishment.

RIGHT Aftermath of a raid by Hitler's revenge weapons: a police officer helps an injured woman from the wreckage of her East London home in 1944.

OPPOSITE BOTTOM A V2 blasts off from a test stand at Peenemünde.

Francis Gary Powers

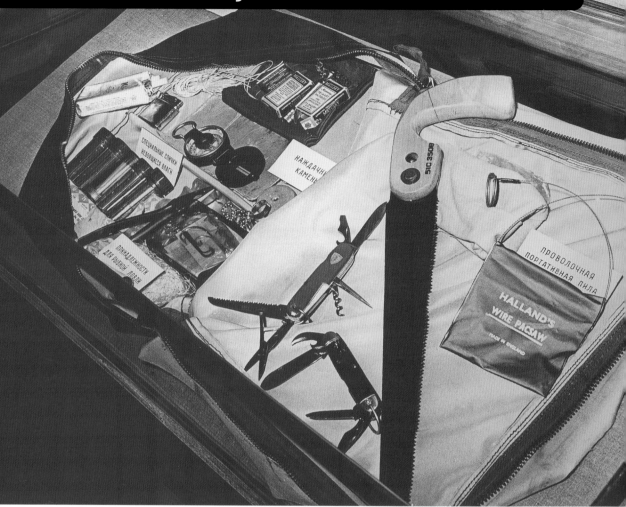

ABOVE The emergency escape and evasion kit carried by the American U2 spyplane pilot Francis Gary Powers, including fishing gear, a wire saw and moisture-resistant matches.

After four fruitful years of flights over Soviet territory, the U2 spyplane's immunity to air defense systems came to an end on May 1, 1960. At 6:26 in the morning, a U2 lifted off from a military airfield near Peshawar, in Pakistan, for the first south-north transit of the Soviet Union, a 3,800-mile mission that was intended to end with the aircraft touching down at Bodo, in Norway. Its flightpath would carry it over Afghanistan, past the Tyuratam Cosmodrome to monitor any space launches that might have been timed to coincide with the May Day parade in Moscow, then across the Sverdlovsk industrial region and the Plesetsk missile test complex, before crossing the Arctic coast of the Soviet Union over the high-security naval and missile sites near Murmansk. Finally, it would skirt the northern coast of Sweden until it reached its destination.

There had been one or two problems, including a 20-minute delay in the takeoff clearance and an intermittent fault in the

КОМАНДНАЯ САМОЛЕТНАЯ
РАДИОСТАНЦИЯ AN/ARC-34
ПРЕДНАЗНАЧЕНА ДЛЯ СВЯЗИ С ДРУГИМИ
САМОЛЕТАМИ И С НАЗЕМНЫМИ РАДИО-
СВЯЗНЫМИ СТАНЦИЯМИ В ПОЛЕТЕ
ФИРМА МАГНОВОКС

UHF COMMAND RADIO
AN/ARC-34
A RADIO SET USED TO RECEIVE OR
GIVE COMMANDS, AS BETWEEN ONE
AIRCRAFT AND ANOTHER OR BETWEEN
AN AIRCRAFT AND THE GROUND

MAGNAVOX CORP

TOP Powers leans on the rail of the dock during his trial in the Moscow courtroom.

ABOVE Items in a special Soviet spyplane exhibition at Moscow's Gorki Park included the U2's command radio for contacting ground stations and other aircraft, and the pilot's list of radio frequencies.

RIGHT The wreckage of Powers' aircraft at the Gorki Park exhibition.

autopilot. In previous weeks, Soviet protests at these flights, backed up by accurate records of the routes across their territory, showed that the aircraft had been tracked by Russian radar systems. Furthermore, on earlier missions, pilots had occasionally seen fighters and anti-aircraft missiles trying, and failing, to reach their high altitudes. However, the pilot of this mission, Francis Gary Powers, settled into the routine of following his precisely specified flightpath as accurately as possible. He had passed the halfway point of his long flight, having flown over Sverdlovsk, when the sudden impact of a Soviet anti-aircraft missile blew his world to pieces. His aircraft plunged into a steepening dive.

By the time that Powers struggled free of the wreckage, he had plummeted to a height of just 15,000 feet, where a barometric device opened his parachute automatically. Within minutes, he made a soft landing in open country, but by the end of the day he was locked in a cell at Moscow's Lubyanka prison. In a little over three months, he would be tried as a spy in Moscow and given a long prison sentence. Fragments of his plane were put on show in the capital, and the Soviets made the most of this valuable propaganda coup. A forthcoming summit meeting in Paris, between Soviet leader Nikita Khrushchev and President Eisenhower, scheduled for May 16, was summarily canceled, and an existing invitation for the President to visit Russia was withdrawn. It seemed that the damage to the relationship between the superpowers would be deep and long-lasting.

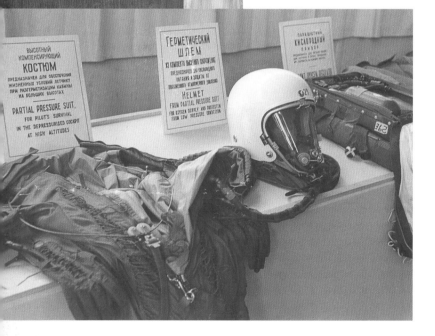

The Cuban Missile Crisis

5 TRUCKS UNDER
CAMOUFLAGE NETTING

Following the Powers disaster, the U2s were restricted to specific shorter-range missions, well away from Soviet airspace. Yet a little more than two years after Powers had parachuted into Russian captivity, these remarkable aircraft would prove their true value in helping to avert a crisis that could have triggered World War III, with devastating consequences for the entire planet.

During the late summer of 1962, one of the regular operations for U2s belonging to the US Air Force's Strategic Air Command, based at Laughlin Air Force Base in Texas, was a fortnightly mission over Cuba. Usually, the prints showed nothing particularly sensitive, but after a mission on August 29, 1962, they revealed new construction work at a number of sites, which were clearly identified as surface-to-air missile bases.

In themselves, these posed no direct threat to the USA, only 150 miles to the north. But when the photographic interpreters reviewed material shot more than two years earlier over the Soviet Union, they found pictures that showed all too clearly identical SAM sites protecting a much more worrying feature—a ballistic missile launch site. Could the new constructions in Cuba be intended to perform the same role, as part of a system that would pose a deadly threat to the United States, bypassing all its early-warning systems, which were designed to monitor inbound missiles from Soviet territory?

Immediately, U2 flights over Cuba were doubled, but no trace of ballistic missiles was found. Yet agents operating within Cuba reported long cylindrical objects, shrouded in tarpaulins, arriving as deck cargo on Russian freighters. The tension mounted over the ensuing weeks until at last, on October 14,

BACKGROUND Aerial photographs provided US President John F. Kennedy with the evidence that Russia was installing nuclear ballistic missiles at launch sites on the island of Cuba in October 1962.

BELOW A Soviet merchant ship loaded with deck cargo, photographed by a US reconnaissance aircraft while on passage to Cuba in October 1962.

1962, two more U2 missions brought back pictures that clearly showed the construction of the first ballistic missile launch complex on Cuban soil.

This was a moment of immense danger in the precarious nuclear standoff between the superpowers. The Kennedy administration ordered an all-out effort to confirm the size and nature of the threat. By October 20, American aircraft had identified the positions and courses of more than 2,000 ships all over the Atlantic Ocean that might conceivably have been en route to Cuban ports. In the meantime, more U2 flights, backed up by low-level photo-reconnaissance missions by US Air Force and US Navy aircraft, had revealed a total of four ballistic missile sites under construction, with 22 of the 24 SAM sites so far identified already operational.

On October 22, the United States announced an embargo on the shipping of missiles and missile equipment to Cuba. The US Navy would stop and examine all vessels bound for the island to inspect their cargoes. A further announcement was made that the launch of any missile from Cuba against any nation in the Western Hemisphere would be regarded as a Soviet nuclear attack on the USA, leading to a full retaliatory strike on Soviet soil.

On October 27, when there had still been no reply from the Soviet leadership, a U2 was shot down by a SAM fired from Cuba. The world teetered on the very brink of war, and US forces were placed on maximum readiness. Then, on the following morning, a message was received from Khruschev, confirming that the Russian

BELOW Another aspect of the Soviet arms buildup in the Western Hemisphere: this US reconnaissance photograph shows a Cuban airfield with 21 Ilyushin Il-28 bombers parked around the perimeter.

BOTTOM US photo-reconnaissance aircraft bring back more pictures of Cuban missile sites.

LAUNCH PAD WITH ERECTOR

LAUNCH PAD WITH ERECTOR

MISSILE READY BLDGS

CABLING

OXIDIZER VEHICLES

FUELING VEHICLES

LAUNCH SITE 1

WA

VEHICLE REVETMENTS

CONTROL BUNKER

CONTROL BUNKER

LAUNCH PADS

PRE-FAB CO

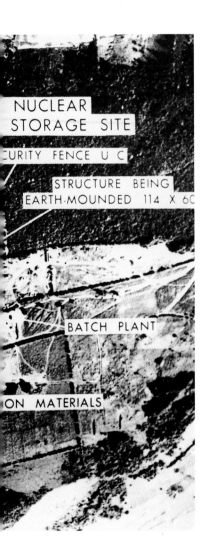

NUCLEAR
STORAGE SITE

CURITY FENCE U C

STRUCTURE BEING
EARTH-MOUNDED 114 X 6C

BATCH PLANT

ON MATERIALS

missile sites on Cuba would be disassembled and the rockets would be shipped back to Russia. To provide confirmation, two US Air Force RF101 Voodoo reconnaissance planes were sent in on a low-level mission to photograph the missile sites. Dodging anti-aircraft fire, both aircraft completed their mission safely, and the prints they brought back confirmed what everyone was praying to see. The missiles were being disassembled, and soon naval reconnaissance forces reported Russian freighters sailing back to their home ports with deck cargoes of missiles. The world's most dangerous crisis had been discovered and averted with the aid of reconnaissance aircraft.

LEFT US reconnaissance photograph shows an intermediate ballistic missile site under construction in Cuba. When operational these missiles would have a range of 2,200 miles.

BELOW US photo-reconnaissance aircraft show a further 17 of the 2,000-mile-range Ilyushin Il-28 bombers still packed in their shipping crates.

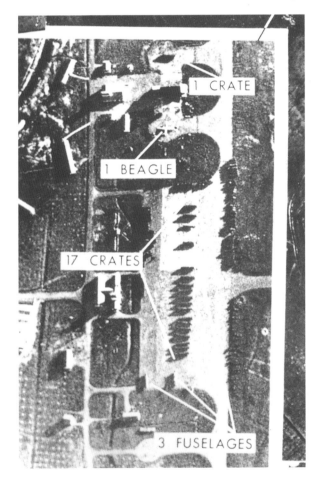

1 CRATE

1 BEAGLE

17 CRATES

3 FUSELAGES

LEFT Oleg Penkovskiy on trial in Moscow for his part in the Cuban missile crisis. His undercover work for the British and Americans helped to resolve the conflict.

171

Satellites and Space Intelligence

ABOVE Spy satellites provided top-quality pictures of territory that had not been covered previously, revealing information like the locations of military airfields, without the risk of attack by hostile aircraft or missiles.

OPPOSITE Within 24 hours of the announcement of the launch of the second Russian satellite, an animated movie shown in London was said to reveal the secrets of the launch of the original Sputnik, including the separation of one of the booster stages (left of picture) from the section containing the satellite (right of picture).

From the beginnings of aerial reconnaissance, the search for valuable intelligence has resulted in a struggle between the pilots of the aircraft involved and the defenses, in the form of fighters, anti-aircraft artillery, and surface-to-air missiles. Where reconnaissance planes try to spy on a country's military, industrial, and communications resources, the bullets, shells, and missiles fired by the defenses are intended to deter them from approaching their targets or, if they succeed in this objective, to ensure that the aircraft do not return to base with their information.

International laws recognize each nation's sovereignty over its airspace in the same way that it has sovereignty over its land territory and its coastal waters. Any foreign aircraft that enters this airspace without authorization risks being driven off, being ordered to land or—in extreme circumstances—being shot down. At the very least, an intrusion into another country's airspace risks causing a major international incident, which can rupture relations between otherwise friendly powers for years.

Yet since October 1957, there has been a means of reconnaissance that enables countries with the necessary technology to obtain a very detailed view of every part of another country's territory without risking any kind of retaliatory action at all. With the launching of the Soviet satellite Sputnik 1, a small metallic sphere weighing 184 pounds, the world of large-scale intelligence gathering was changed forever.

The tiny satellite was blasted into space by a multi-stage rocket containing no fewer than 24 rocket motors, which pushed it to a final speed of 18,000 miles per hour. This left it in earth orbit hundreds of miles above the surface, far beyond any legal concepts of national airspace. And although it was loaded with a miniature radio transmitter and batteries, rather than an array of cameras, the precedent had been established. In the years that followed, satellites and their capabilities would expand dramatically, providing their builders with an unhindered view into every corner of their neighbors' and adversaries' backyards.

The first intelligence satellites

Although the early years of the international space race were dominated by the Soviet Union, thanks mainly to the power and reliability of its rockets, which were able to put progressively heavier payloads into orbit, the American satellites carried increasingly sophisticated instrumentation. Furthermore,

as the ability to place a satellite in orbit as a benchmark of superpower technology began to fade, the usefulness of these high-altitude platforms became more obvious.

For example, the early satellites tended to have orbits that caused them to pass over most parts of the earth's surface in successive circuits. More powerful rockets sent them into higher orbits, giving them a wider "footprint," or area of coverage, at the expense of being farther away from the surface for observational purposes. On the other hand, more accurate launch techniques enabled some satellites to be placed in geosynchronous orbits. In these orbits, the speed of the satellite through space exactly matches the rotation of the earth below, so that it remains over the same spot on the surface. This allowed satellites to be used as navigational beacons, as communications relays, and as early-warning devices to monitor certain types of activity, such as missile launches and similar sensitive events.

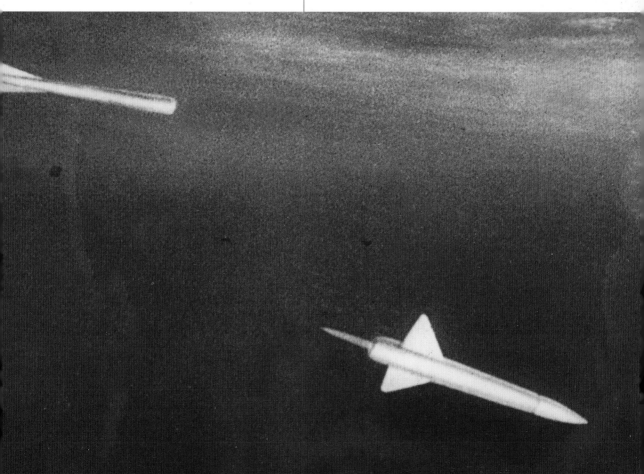

Cameras in the sky

Although the American satellite launch program began behind that of the Soviets, it soon made up much of the lost ground. On April 1, 1960, a Thor-Able rocket adapted from a ballistic missile blasted off from Cape Canaveral in Florida, to put a meteorological satellite named *Tiros-1* (for Television and InfraRed Observation Satellite) into an elliptical earth orbit, varying in height from 431 miles to 471 miles above the surface. The satellite weighed a total of 290 pounds, and was covered with some 9,000 small solar cells to generate the energy needed to replenish its internal batteries. Inside were two small TV cameras that relayed their pictures directly to ground receiving stations when the satellite was in range, and recorded them on tape when out of range, to be relayed later in its orbit.

The results were a revelation. Even these early cameras showed a clear view of airfields, missile sites, and other defensive installations all over the world. Equally clear was that dedicated intelligence satellites could be developed to provide much more detail of the ground over which they passed, and less than two months later, on May 24, 1960 (the same month as Gary Powers' fatal U2 mission—see chapter 8), the first US reconnaissance satellite was launched into orbit. This was *Midas* (for Missile Defense Alarm System), which was equipped with sensitive infrared detectors to pinpoint the plume of heat energy emitted when a ballistic missile was launched from any point on the earth's surface.

Midas was followed by a program called Corona, using satellites in the Discoverer series, *Discoverer 13* being launched on August 10, 1960. These satellites employed conventional film cameras rather than electronic devices to produce pictures with finer resolution, which raised the problem of retrieving the exposed film without bringing the satellite back to earth. A system was developed that allowed the satellite to eject a capsule of exposed film when in the right position, in a container fitted with a parachute and a homing beacon, so that it could be retrieved at low level by specially equipped aircraft, but persuading the system to work properly had taken 18 months and the launch of 11 Discoverer satellites.

In a trial on August 11, 1961, *Discoverer 13* ejected an empty capsule. The aircraft that was supposed to capture the capsule's parachute missed its target, allowing the capsule to fall into the ocean, where it floated until picked up. However, this proved that the system worked. A week later, *Discoverer 14* was launched with the full film and camera system. Its orbit took it over the whole of the

RIGHT The first satellite ever to have been recovered from space was the American *Discoverer 13*, which was retrieved from the sea off Hawaii and flown to Washington to be shown to the President.

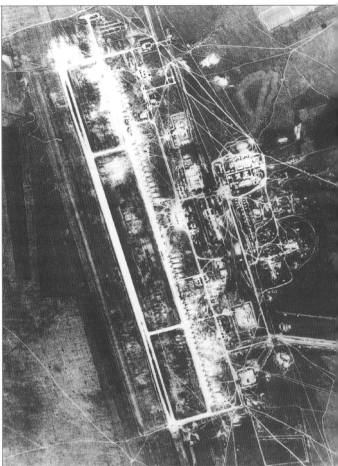

Soviet Union, and when the exposed film was recovered, the pictures showed airbases and rocket sites, albeit with a fairly crude resolution of between 50 and 100 feet. A new age of espionage had begun.

The truth about the missile gap

During 1961, further Discoverer satellites showed the first Soviet ICBM site and test facility at Plesetsk in North-Western Russia, in a layout that was similar to the space research launch site at Tyuratam. Additional flights provided more detail, but their most staggering achievement was to refute the concept of the missile gap, which had caused many Western intelligence agencies to believe, in the absence of any hard information, that the Soviet Union had established a huge advantage in the number of ballistic missiles deployed. The reality was very different. The frontline Soviet missile, the SS6, was large and heavy, and it could only be moved on railroad flatcars or big trucks that needed major high-quality highways. All these transport links were clearly shown on the satellite

ABOVE LEFT This satellite photograph of a Soviet SS-9 ICBM launch silo was one of several spy pictures released by US intelligence officials in May 1995.

ABOVE Another American satellite photograph, recently declassified and placed on the internet, shows a large Soviet long-range bomber base near Dolon in Kazakhstan.

photographs, without any evidence of missiles to match Khruschev's bombastic and bloodcurdling boasts.

Although the Americans had overtaken the Russians with their intelligence satellite program, the Soviet Union was able to reply with *Cosmos 4*, launched on April 26, 1962, in an orbit that allowed it to cover the whole land area of the USA. This remained in orbit for only three days, before returning to earth with all the information it had been able to gather. During the next two years, a series of

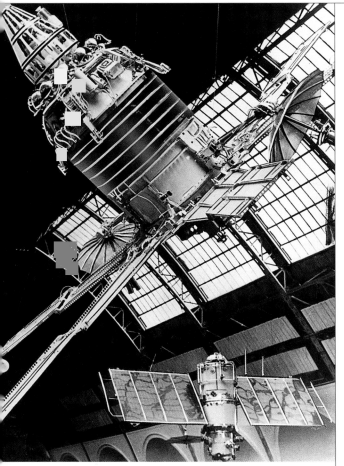

ABOVE A Soviet long-range intelligence gathering satellite, Molnija I, assembled in the Space Pavilion at the Moscow People's Economic Exhibition of 1967.

Cosmos satellites was launched, many of which had an intelligence gathering role. They differed from the US satellites in one important respect, however: they tended to be launched into higher orbits, which enabled them to stay aloft for longer periods, but which limited the resolution of the pictures they were able to obtain.

The Americans followed up with a whole succession of satellites that monitored the electronic signatures of Soviet, Chinese, and other air defense and missile control radars. Another series of satellites, under what was named the Gambit program, was designed to switch from the broad coverage of the original photo-reconnaissance satellites to close-up images of specific areas of interest.

Total coverage—and new hope for the future

By 1963, the KH7 satellites were following very low orbits only 70–80 miles above the surface, which enabled them to produce pictures with a resolution of only 18 inches. At the end of that year, the CIA was able to report that satellite coverage gave a clear view of all the Soviet Union's large cities, all but one of the major submarine bases, all of the heavy bomber bases, a large proportion of the country's railroad network, and all of the known ICBM complexes. At the same time, the Soviet satellites were monitoring all the key military sites of the United States and the Western allies, to record details of the readiness of different missile systems and launch sites, to watch naval and military operations, and to monitor tests of new weapons.

Developments during the 1970s, 1980s, and 1990s allowed satellites to produce vast amounts of progressively more detailed material. Eventually, they would be able to turn intelligence gathering from an activity aimed against real or potential adversaries into a system that worked for the benefit of all, to create an infinitely safer world. To begin with, increasingly powerful and sensitive cameras, and progressively greater areas of coverage, made it impossible to hide the construction of missile launch sites. This made it almost certain that a country's defenses would be known and precisely targeted by the missile forces of the other side, reducing the likelihood of even the most powerful nations being tempted to risk a first strike.

However, in case the unthinkable happened, the United States' Minuteman concept was developed to ensure that some of the country's missile resources would be able to survive a surprise attack, and be capable of delivering a

OPPOSITE The Chinese nuclear test site near Lop Nur was monitored by US satellites during its construction in the 1960s. This picture was taken by a KH4 spy satellite on October 20, 1994, four days after China's first nuclear test.

retaliatory strike. Far more launch sites were built than the number of missiles available, on the assumption that Russia would have to hit every one to be sure of delivering a successful pre-emptive strike. Only a proportion of the sites would contain missiles at any one time, but there would be nothing to allow satellite observation to identify them, and it was easier and cheaper to build launch sites than it was for the Soviets to increase their missiles by a corresponding number.

However, as continuing East-West negotiations strove to develop a series of arms-limitation and arms-reduction treaties, the role of the satellites became more positive. Earlier negotiations had often stalled because of the difficulty of monitoring compliance with the conditions of the treaties to the satisfaction of all parties. Access to missile sites was barred on security grounds, but once the satellites were able to monitor changes from space, this stumbling block was removed. With mutually agreed limits on missile deployment, on anti-missile defense systems, and on conventional force levels, the framework was in place for the ultimate in intelligence systems to help produce an end to the arms race, and a limitation on nuclear overkill.

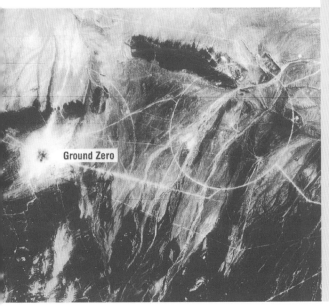

Ground Zero

Star Wars

Apart from gathering normal intelligence, satellites had an even more crucial role to play in the original "Star Wars" Strategic Defense Initiative ballistic missile defense system first proposed in the 1980s. Satellites were to warn of the launch of hostile missiles, and carry many of the essential parts of the system, including the lasers for destroying the approaching missiles and warheads, and the focusing mirrors for directing the laser beams onto their targets.

Lasers were proposed as the means of destroying hostile missiles because bursts of laser energy travel at the speed of light, so in theory their responses would be fast enough to hit incoming missiles. In addition, it is possible to keep laser beams very narrowly focused over a long distance using mirrors, and they can carry a lot of energy. Thus, the high-energy lasers would burn holes in the skins of ballistic missiles so that they would weaken and split, destroying the missiles.

Unfortunately, the chief problem with making the system work was one of scale. With modern ballistic missiles able to split into a series of independently targeted warheads, the defenses would be faced with a huge number of targets in a short space of time, a difficulty compounded by the possibility of the enemy using decoy warheads. At the time, intelligence estimates suggested that the Soviet missile forces could have launched as many as 100,000 independently targeted nuclear warheads at once, which would have posed an incredibly difficult threat for the defensive system to overcome.

Another potential problem arose from the fact that it would have been essential to focus a laser on the same spot on the skin of a fast-moving missile or warhead to burn through it. Hardening the missiles by adding heatshields, or simply spinning them in flight, would have greatly increased the time needed to achieve this. In addition, very large, high-quality mirrors would have been needed to focus the lasers, and even the tiniest imperfection of a fraction of a micron on a mirror's surface would have reduced its effectiveness considerably.

Furthermore, the mirrors would have had to survive being launched into orbit, would have needed to maintain exactly the right attitude in space, and would have had to withstand a laser beam powerful enough to destroy a missile. In all, more than 100 orbiting mirrors would have been needed to aim the laser beams. Finally, each laser would have had to be around 100 times more powerful than existing military lasers, and the mirrors would have had to be capable of switching between targets accurately and instantly. The laser beams would have had to be perfectly aimed, and targeting decisions taken instantaneously, all of which added up to a formidable assignment, even for the technology of the world's most capable and sophisticated satellite specialists.

Nuclear emergency

ABOVE An aerial view of the wrecked reactor block at the Chernobyl nuclear power plant, which leaked a cloud of radioactive gas that spread across Europe in April 1986.

Not all satellite missions relate to military emergencies. On August 28, 1985, the United States was launching a second KH11 satellite to replace one that had been brought down to earth a month before. However, only minutes after the launch, the motors of the Titan rocket failed, and ground control had to use the onboard self-destruct mechanism to blow both rocket and satellite to pieces before they could crash back to earth. This disaster left only one KH11 in orbit, and when the Americans attempted to launch a KH9 satellite on April 18, 1986, using another Titan rocket, this too ended in failure due to an explosion in one of the boosters, which activated the self-destruct system and blew the rocket apart.

As a result, the United States was badly placed to respond immediately when a major emergency occurred at the Chernobyl nuclear power plant, 80 miles from Kiev in the Ukraine. Technicians at the plant had been practicing safety drills on April 26, 1986, but had been using non-standard procedures that resulted in a runaway reaction, which blew the reactor apart and released huge quantities of radioactive material into the atmosphere.

The first hints of a crisis had been suggested on the next day by an increase in communications traffic in the area. This was followed a day later by the official Soviet statement.

Nuclear monitoring aircraft, based at Mildenhall in Britain, were sent to fly over Europe and the Mediterranean to measure the fallout, while the sole remaining KH11 was set to check out the area later that same day. The orbit it was following at the time gave only a distant view, which produced little definite information, but by evening on the following day, April 29, the third orbit provided detailed pictures from directly over the

shattered reactor, showing that the roof was missing and the walls had been blown out, leaving the interior a glowing mass of radioactive material.

When data tapes of the US Air Force's missile launch monitoring satellites were studied, they showed infrared images of a sudden explosion, which was the blast that had blown the roof off the reactor. Further orbits of the KH11 produced continued coverage of the cleanup operation, showing that the fire had been damped down, and that helicopters were hovering overhead and dropping sacks of sand into the reactor interior to seal it off from the atmosphere. These were followed by tons of lead pellets to fill the gaps between the sacks and provide a radioactive shield. The satellite pictures also showed that the other Chernobyl reactors seemed unaffected by the catastrophe, something that had caused a great deal of concern in the absence of precise information from the Soviet authorities.

ABOVE A worker checks levels of radioactivity near the damaged reactor block at Chernobyl.

BACKGROUND This aerial view of the Chernobyl nuclear power plant shows the site of the original accident, which killed four of the plant's workers, according to Soviet sources.

Prelude to the Gulf War

BELOW US satellite imagery, showing evidence of Iraqi troop movements on the Kuwaiti border as a prelude to invasion, July 1990.

The first signs of Saddam Hussein's impending invasion of Kuwait were revealed by United States satellite surveillance on July 16, 1990, when photographs showed a brigade of Iraqi T72 tanks occupying a previously empty sector of desert in the south-east of the country, close to the Kuwaiti border. A day later, more tanks had appeared, increasing the strength to two divisions, and on July 18, satellite pictures showed a third armored division alongside the first two. The details and the repeated coverage that showed the speed of the buildup were characteristic of first-class satellite intelligence, and they set alarm bells ringing among the US intelligence community.

Troop
Encampment

There was a possibility that these large-scale troop movements were intended simply as a threat, to force Kuwait to act in a manner that was more acceptable to its large and bellicose neighbor. However, ongoing satellite coverage showed the military buildup continuing, beyond the point at which it became clear that a full-scale invasion was almost certain within days. By August 1, US intelligence predicted an attack within 24 hours. On the morning of August 2, 1990, Iraqi troops proved that estimate correct with perfect timing, crossing the border into Kuwait.

As the furious diplomatic offensive in the West gathered pace, to assemble a coalition of friendly forces in the region to pressurize Iraq into withdrawing to its own territory, and to defend Saudi Arabia from Iraqi forces, the satellite resources available to the United States were of vital importance. In addition to three orthodox KH11s, one of which had been launched in 1984, there was an advanced KH11, which had an infrared imaging capability to produce pictures at night. There was also a Lacrosse radar equipped satellite, which could produce less detailed images even through cloud in conditions of complete darkness.

ABOVE Iraqi troops take cover behind their armored personnel carrier as Kuwaiti resistance snipers open fire on them in the Salmya district, during the Iraqi invasion of Kuwait on August 2, 1990.

BELOW An Iraqi tank moves through the streets of Kuwait on the morning of the invasion.

This battery of high-tech satellites was reinforced by specialized aircraft flying close to Iraqi airspace, monitoring the smallest details of the Iraqi deployments on the ground. At the same time, the Iraqis had no means of spotting the forces being assembled against them. They had no satellites, or access to satellite information, and reconnaissance aircraft were shot down by coalition fighters whenever they tried to operate.

As a result, the coalition force commanders enjoyed a priceless advantage over their Iraqi opposite numbers, when they took the offensive to liberate Kuwait in January 1991. Nevertheless, the situation revealed vital gaps in the satellites' coverage. With the termination of the KH9 program, there was no means of keeping the new generation of satellites over the area of the conflict for more than a short period each day. Ironically, another problem was created by the deadly accuracy of the "smart" weapons used by the coalition air forces to attack Iraqi buildings, command bunkers, and other installations. Because these bombs and missiles were designed to penetrate the outer walls of buildings and explode inside to cause maximum destruction, the only signs they betrayed to even high-precision satellite surveillance were the holes through which they had entered their targets, with no evidence of the actual damage inflicted.

Space Wars

With satellites proving to be such a powerful inhibitor of a nation's warlike moves, removing the threat they presented was bound to be a first step in any genuine attempt to attack the other side in an East-West nuclear exchange. To this end, the Soviet Union experimented with a number of different anti-satellite weapons systems.

In early 1971, a series of practice interceptions was conducted by Russian satellites. The Soviet *Cosmos 394* satellite was launched on February 9, 1971, and was used to destroy *Cosmos 400*, which had been launched on February 25. In the following month, the routine was repeated. *Cosmos 400*, launched on March 19, 1971, was seen to disappear after a close approach by *Cosmos 404* on April 3, at an altitude of about 550 miles. *Cosmos 462* was launched on December 3, 1971, and was destroyed by *Cosmos 459* in a much lower orbit, at a height of only 143 miles.

Less clear were the methods that had been used to achieve this destruction. In some cases, both satellites appear to have been destroyed, suggesting a suicide-weapon satellite that was steered as close as possible to its intended target before explosives aboard were detonated by a signal from the ground, blowing both satellites to pieces. In other cases, the hunter-killer satellite probably launched a cloud of small metal pellets into its target's path which, given the vulnerability of satellites and their high closing speed, either would have destroyed the target completely, or wrecked its onboard systems and rendered it useless.

The disadvantage of these weapons-carrying satellites was that they were inherently cumbersome, needing powerful rockets to launch them into orbit. Furthermore, they needed time for ground control to move them into an attacking position, during which their purpose would be increasingly obvious to the other side's surveillance and defensive systems.

In contrast to the Russian approach, the American anti-satellite systems were based on missiles aimed directly at their targets. The early devices were surface-to-space missiles, but in the 1980s the Reagan administration sponsored the development of a smaller missile called an MHV (miniature homing vehicle). This was air-launched from a modified F15 fighter, and with accurate sensors the missile was capable of scoring a direct hit to destroy its target, so no explosives were needed, and there was a possibility that the nature of the attack could be kept secret.

The missile could be launched more than 10 miles above the earth, and as the F15 was a carrier-borne fighter, it could be used almost anywhere in the world. MHV's disadvantage, however, was that it could only hit satellites in relatively low orbits.

OPPOSITE A satellite of the Soviet Cosmos series.

BELOW A Cosmos satellite is launched from the Plesetsk Cosmodrome in the Arkhangelsk region of northern Russia.

Espionage in the Twenty-first Century

ABOVE An artist's impression of the Aurora spyplane high in the atmosphere—believed by many sources to be a top-secret design to replace the extremely successful US SR-71 Blackbird surveillance aircraft.

OPPOSITE Russian guards stationed outside their embassy in Washington DC following the expulsion of four suspected Russian spies from the United States on March 22, 2001.

In many ways, the importance of espionage seems to have reached its peak in the twentieth century. There are several reasons for this. It was in the twentieth century that real or potential conflicts became more destructive and terrifying than ever before. In earlier times, wars were often fought by small professional armies over obscure questions of sovereignty or dynastic succession, and the use of spies to steal an advantage, while a worthwhile challenge to those involved, had relatively little effect on the course of history. But the terrible world wars of 60 and 80 years ago posed much greater threats to national survival, and as a result any and every possibility of ferreting out an enemy's secrets, and of planting false information, offered greater benefits than had ever been known.

Furthermore, the increasingly uneasy peace after 1945, with two superpowers armed to the teeth with weapons that were capable of destroying all life on the planet several times over, raised the espionage stakes to an almost impossibly high level. In addition to the vital contribution that still could be made by devoted and courageous human agents, there was a battery of new technologies that could be used to divine a potential enemy's intentions and resources, and to mislead that same potential enemy about the plans and capabilities of one's own side. Spyplanes and satellites, electronic and communications intelligence, traffic analysis and cipher breaking, and a whole range of subsidiary developments made the intelligence and information picture more complex and challenging than it had ever been.

The end of the Cold War

Ironically, these dramatic developments were perfected at almost exactly the time when the old Cold War certainties were about to dissolve forever. In a few short, but dramatic, years in the early 1990s, the awesome threat still posed by the deployment and the weaponry of the Soviet Union and Warsaw Pact countries crumbled and disappeared. From their positions on the far side of the River Elbe in central Germany, the frontline units of the Russian Army withdrew hundreds of miles, to beyond the far borders of newly independent fragments of the Soviet state, like Belarus. Old republics in the Baltic, in Central Asia, and even in the Soviet heartlands, like the Ukraine, regained long-forgotten independence.

For a time, it seemed that the world had become a much safer place. The old secrecy vanished along with the Iron Curtain fortifications of watchtowers and deathstrips. Suddenly it was possible to talk, almost without restriction, to agents and chiefs of the KGB, to visit their museum in the Lubyanka, and to hear at firsthand the opposition's view of hostile espionage operations, which had been maintained under a suffocating cloak of secrecy when they had been mounted. In this new climate of openness, there was a possibility that espionage itself might become as dated as the barbed wire and minefields that had divided Germany.

Yet if such a possibility ever seriously existed, it soon vanished in the glare of the post-Cold War future. Perhaps the old threats really had disappeared, but one legacy of their passing was a degree of suspicion between East and West that not all their interests coincided completely at the start of the twenty-first century. Russian fears about the eastward expansion of NATO, resulting from their former Warsaw Pact allies, like Hungary and Poland, lining up for membership of the one organization they felt could guarantee them safety from a resurgence of Russian power in the future, were matched by Western worries about newly democratic Russia sliding back into totalitarianism and militarism. Even if the old hostility was missing, there were plenty of good reasons for keeping a close watch on such powerful friends, as there had been on the old enemies, and many of the Cold War agencies and networks continued as if nothing had changed.

New links between old adversaries

Early in 2001, in a scenario that would have been familiar in the 1940s and 1950s, the United States ordered the expulsion of 50 Russian diplomats suspected of being espionage agents. Since the appointment of Vladimir Putin, a career KGB officer, as President of Russia, espionage efforts have increased in importance. During the previous year, Sergei Nikolayevich Lebedev was appointed as the new head of the Russian Foreign Intelligence Agency, the SVR. Lebedev had previously been in charge of the SVR station in New York, and presumably was an expert on Russian espionage networks in the USA.

Experience has shown two factors operating in cases of mass expulsion. Because the usual reaction of one country to another expelling a number of its diplomats is to deport the same number of that country's representatives, by way of retaliation and as a means of suggesting that both sides are equally guilty of clandestine intelligence gathering, these mass expulsions are rarely undertaken lightly. Secondly, they are immensely damaging to one or both sides because of the loss of experienced and well-established agents, and it may take years to rebuild those lost networks. For the Americans to risk harming US-Russian relations by such a precipitate action, which might well have led to their own networks within Russia being compromised, there must have been a clearly perceived threat to US national security.

However, the most extraordinary fact about this latest expulsion is the background against which it was carried out. The SVR has regular, and often convivial, contacts with its opposite numbers in the CIA, and the two services regularly collaborate on operations in areas of mutual interest, like the suppression of international terrorist threats for example. This action represents a sharp reminder that those interests do not always coincide, and that Russia, with its rising crime and declining economy, allied to a still mighty nuclear arsenal, remains a major target for Western intelligence gathering.

New threats for new technology

The biggest change to international espionage, in addition to the ending of the Cold War, has been the runaway proliferation of states that now have nuclear and biological capabilities, and in many cases the ideology and objectives that may persuade them to use this power to blackmail much larger and more powerful countries. India and Pakistan have

raised the stakes with successful nuclear tests and increased levels of armaments. Iraq has been successful in expelling UN weapons inspectors, and in concealing the level of its research and development of nuclear, chemical, and biological weapons. North Korea, for the present, remains outside the influence of any power bloc, although severe economic problems have resulted in a lessening of tension with its southern neighbor. And China, as always, remains an increasingly powerful and increasingly confident enigma to the current global power

ABOVE Into the nuclear age: an Indian missile with nuclear warhead takes part in a military parade; both India and Pakistan have developed atomic weapons.

structure, which calls for a close watch on its military and political capability.

In some respects, the latest technology raises hopes that these multiple threats can be kept under close and continuous surveillance. Yet the astonishing capabilities of modern satellites and sensors still suffer from limitations. During the Gulf War, units of the Iraqi Army were able to decoy coalition aircraft into expending scarce and expensive missiles by setting up inflatable decoys fitted with heat sources to produce the right kind of signature to infrared target seeking systems. And the extensive satellite coverage of the war zone was still unable to pinpoint the Iraqis' mobile launchers for their Scud surface-to-surface missiles which, when fired at targets inside Israel, nearly widened the conflict and broke the cohesion of the coalition.

On the other hand, the Iraqis were completely confused by the coalition's use of radio-controlled pilotless drones, which drew the fire of the Iraqi air defense system, to the point where genuine attacks sent in later encountered much less opposition. This was a specialized use of drones, which represent another major step forward in intelligence gathering over relatively small areas at low level, since they are small, cheap, and expendable, and do not show up well on radar. Whether launched from the ground, from ships or from aircraft, they can loiter over a target area and send back visual, electronic or infrared information for a much longer period than a satellite.

TOP The Pakistani nuclear test site, in the remote Chagai district of the province of Baluchistan, as its fifth nuclear bomb was detonated in an underground test.

RIGHT India's nuclear test site in the province of Rajasthan, not far from the border with Pakistan, where five underground nuclear tests were carried out.

A change of target?

If there is a challenge to this brave new world of technological espionage, it probably lies in the political arena, and in the West rather than the East. The European Union has been growing increasingly restive at the long-established UK-USA joint agreement on intelligence sharing, which originated in 1942 as part of the Anglo-American alliance of World War II, and which also involves the intelligence agencies of Canada, Australia, and New Zealand. In particular, the Echelon electronic intelligence system, which filters information gathered by Britain's GCHQ and America's CIA, has been accused of eavesdropping on Europe's fax, e-mail, telephone, and satellite transmissions, using listening stations on British soil, like Menwith Hill in Yorkshire.

France in particular has suggested that British participation in this program may be in breach of Article 8 of the European Convention on Human Rights, which guarantees privacy, and Article 10 of the Treaty of Amsterdam, which commits member states to uphold common EU interests.

On the other hand, British and American sources insist that the only commercial use of the program has been to expose attempts by European companies (French in at least two cases) to use bribery to try to win defense contracts, and to infiltrate US multinationals. Similarly, Germany has targeted companies in Italy, France, the US, and the UK to gain commercial advantage. Furthermore, there has been no suggestion by other European countries that they are willing to disassemble their own intelligence organizations in favor of a Europe-wide intelligence system, nor have they offered to open them up to EU inspection.

BELOW The Royal Air Force site at Menwith Hill, where Greenpeace supporters staged a rooftop protest in July 2001 against plans for the US "Son of Star Wars" anti-missile defense system.

A bright new espionage world?

Certainly the first years of the twenty-first century have been marked on the American side by two new technological advances. The perceived threat from so-called rogue states is

being met by the development of a new missile defense system, which will be a successor to the original "Star Wars" concept of the 1980s. In an attempt to address international concern about a measure that would appear to negate the existing Anti-Ballistic Missile Treaty, which prohibits such defensive systems as a means of tilting the nuclear balance and destabilizing the equilibrium between East and West, America has proposed cutting its nuclear arsenal to radically low levels.

As this book went to press, newspapers were reporting an apparently bizarre development of this policy, in the form of a suggestion that the defensive shield should use Russian S300 missiles, and that the two nations should hold joint exercises in identifying attacks with these missiles and successfully shooting them down. This could also involve the next generation of Russian missiles, the S400, together with associated Russian radar systems. This proposal seems insufficient to allay Russian fears of the new system, despite the financial benefits arising from the missile sales. Perhaps even more important is the reaction of the Chinese, who are concerned in case a

successful anti-missile defense system negates the technological and financial investments they have made in producing a ballistic missile capability. Just as they are increasing pressure on the USA to cease supporting the regime in Taiwan, and to allow them to restore it to the motherland, they see these new developments as threatening their freedom of action.

The second new US initiative is much less controversial, except on the grounds of cost. According to a report in the Los Angeles Times in March 2001, plans have begun on a $25 billion project called Future Image Architecture, to produce large numbers of new spy satellites for launching from 2005 onward providing tighter coverage of the whole world. These satellites will occupy higher orbits to maximize coverage of an area for twice as long as existing satellites, using high-power optics to make up for their greater altitude, backed up by radar. They will be able to relay thousands of images, zooming in to areas of particular interest, in daylight, darkness or bad weather, and will be difficult to detect because of their smaller size. Having up to two dozen in orbit at any one time would keep gaps in coverage to the bare minimum.

Other technological advances promise to increase the ultimate value of satellite intelligence through the deployment of multiple collection systems. Already the CIA can take images from its latest satellites and process the information through computers to create three-dimensional moving pictures of targets like cities, defense sites, and military installations. Other developments being pursued by the US National Reconnaissance Office, including hyperspectral sensors, would be able to capture images from target objects using a combination of radar, ultraviolet, and infrared sensors, together with conventional optical cameras.

Not only would this give image interpreters a much more detailed picture of the shape, density, temperature, chemical composition, and movement of a target, but it would also greatly increase the difficulty of creating a successful deception. In interpretation terms, it seems that hyperspectral sensors would allow intelligence services to know whether what they see really is what they get. So for the time being, the future of intelligence in the twenty-first century seems as exciting and challenging as ever. On the other hand, on the political, technical, and operational fronts, that future remains as difficult to assess and to predict as ever in the long, colorful and complex history of espionage.

The Ho Chi Minh Trail

During the Vietnam War, one of the most important objectives for the American forces was to cut off supplies being sent to the Viet Cong in the south from Communist North Vietnam along the network of jungle tracks collectively known as the Ho Chi Minh Trail. Because of the impossibility of sending sufficient troops into the jungle to disrupt the flow of food, ammunition, and reinforcements, they decided to use heavy air strikes instead.

However, this required good intelligence to avoid wasting bombs and rockets on empty jungle. To obtain this vital information, aircraft were sent in to seed the whole length of the Trail with huge quantities of remote sensors. Once these hit the ground or became caught up in the jungle canopy, they became active. Some were sonic detectors that could react to noise and movement; some were activated by vibration; and some were triggered by heat sources, or by the chemicals given off by human perspiration. All the signals from the thousands of sensors dropped along the supply routes were picked up and fed to a central computer system across the Thai border, which analyzed where and when supply deliveries were being made. Once the picture had been finalized, American fighter-bombers were sent in to drop huge quantities of napalm and high explosive in the area.

In the end, however, the campaign was only partially successful. Part of the problem was the vast number of different routes that had to be covered, and the large number of small groups rather than substantial convoys that actually carried the supplies. As a result, large and worthwhile targets were rare, and very often a major raid would only cause minimal casualties and disruption to the Viet Cong's supply lines. Another problem was that the enemy developed their own techniques for finding, immobilizing or deceiving the sensors. Sound and movement detectors could be triggered by animals, and even the perspiration sensors could be fooled by bags of human urine hung in the trees.

RIGHT A flight of four US Air Force C-123 "Ranch Hands" spray a suspected Viet Cong position in the Vietnamese jungle with a defoliant liquid to deprive them of cover.

The Walker spy ring

ABOVE John Walker (left), accused ringleader of the Walker spy ring, being led out of a detention center at Rockville, Maryland, on his way to appear at the US District Court in Baltimore. Walker was expected to plead guilty to charges of passing US military secrets to the Russians.

BELOW Fighters, strike aircraft, and patrol planes deployed on the flight deck of the *USS Enterprise* in the Northern Gulf.

I n Russian estimation, their greatest espionage coup against the Americans during the Cold War years, surpassing even the network of agents that had given Stalin the secrets of the atomic bomb, was the spy ring led by a former US Navy officer named John A. Walker, Jr. He had been a Soviet spy since 1968, when he had served as communications watch officer for the Atlantic Fleet submarine command in Norfolk, Virginia. His position gave him access to signals between US submarines in the Atlantic, Arctic, and Mediterranean, and he was able to pass on cipher keys and technical information on US cipher machines to his KGB controllers.

Walker delivered sufficient information on the KL-47 cipher machine for the Russians to make their own version, which they used in conjunction with the keys he had given them to decipher US signals traffic. Later, Walker was transferred to radio school at San Diego, after which he served aboard the carrier *USS Enterprise*, then the *USS Niagara Falls*, followed by a final tour at Atlantic Fleet Headquarters in Norfolk. For eight years, he continued to supply the KGB with cipher keys and technical information on US cipher machines, until he retired from the navy in 1976.

However, by then he had recruited an agent of his own, Senior Chief Radioman Jerry Whitworth, who delivered secret cipher material for another eight years, until Walker was able

LEFT The nuclear aircraft carrier *USS Enterprise* prepares to launch aircraft from her flight deck, while operating in the Northern Gulf.

BELOW The *USS Enterprise* launched a series of strong and sustained air strikes against targets in Iraq when it became clear that Iraqi leader Saddam Hussein was refusing to co-operate with UN weapons inspectors.

to recruit his own brother, Arthur, and his son, Michael. Between them, they turned over a vast amount of material on ciphers, deception, electronic countermeasures, satellite communications systems, and different types of weapon, together with operational textbooks. Michael, in particular, was able to pass on details of highly secret weapons systems, including data on the Tomahawk cruise missile as a result of duty aboard the nuclear aircraft carrier *USS Nimitz*.

According to Soviet double agent Vitaly Yurchenko, Walker's information had allowed the KGB to read more than a million secret US Navy communications. Yet this highly valuable network was finally betrayed because John Walker had revealed his spying activities to his former wife, who had finally alerted the FBI. They were arrested in May 1985, and John and Arthur Walker, together with Jerry Whitworth, were given life sentences, while Michael Walker was sentenced to 25 years in prison.

The Hanssen spy case

Many successful spies have been inspired to work through ideology, a hatred of a particular regime, or a wish to make the world a better place by bringing about a shift in the balance of power. Others, from the beginnings of espionage, have been motivated mostly, or entirely, by money, a trade that is proving increasingly prevalent.

Robert Hanssen, an FBI agent living in Vienna, northern Virginia, worked for the Bureau for 27 years, but in the most recent spy scandal to hit the USA he has pleaded guilty to having worked as a Soviet agent since the mid-1980s for a vast payoff of cash and diamonds worth $1.4 million.

According to a letter written by Hanssen in March 2000, his fascination with espionage began when as a boy he first read a book by Kim Philby, who had worked as a KGB agent for 50 years. In Hanssen's words, "I decided on this course when I was 14 years old." Only the third FBI officer to be accused of spying, he is said to have approached the Russians and offered to provide sensitive counterintelligence information in return for substantial payments.

Hanssen himself was very secretive, preferring to work through dead drops, and never meeting his Russian controllers face to face. Nevertheless, it appears that he passed 6,000 pages of information to them, including details of Russian double agents working for the Americans, leading to at least three of them being recalled to Moscow from the Soviet Embassy in Washington. Upon their return, they were interrogated, and ultimately two were executed, while the third was imprisoned.

Hanssen was arrested in early 2001, when leaving documents for his Russian handlers at a dead letter drop in a suburban park in Virginia. The FBI maintained surveillance on the spot to catch the Russian agents involved, but no one turned up to retrieve the valuable information, leading to worries that there may be more double agents within the FBI, one of whom could have warned the Russians of the trap. As a consequence, the Bureau announced that it would increase the use of compulsory lie-detector tests for employees, and restrict computer access on a more tightly supervised need-to-know basis.

BACKGROUND FBI photograph of the remote, wooded "Lewis" drop used by Hanssen to pass on vital secret information to his Russian spymasters.

ABOVE Police crime-scene tape cordons off the house of accused spy Robert P. Hanssen in Vienna, Virginia, in February 2001. Hanssen was charged with betraying Soviet double agents working for the US, and selling American secrets.

BELOW The charges against Hanssen carry the potential for the death penalty but his guilty plea has had the sentence reduced to a true life sentence where he will have no possibility of parole or early release.

The Chinese spyplane incident

On the first day of April 2001, a four-engined American EP-3 Aries II surveillance aircraft flying over the South China Sea from the Japanese island of Okinawa was involved in a midair collision with a Chinese F-8 fighter, which resulted in the American aircraft having to make an emergency landing at the Chinese airbase at Lingshui on the island of Hainan, where, as this book goes to press, the damaged plane remains under heavy guard.

It was an incident heavy with implications for both sides. The Chinese administration had been growing increasingly concerned at two aspects of current American policy: support for Taiwan, which China regards as a lost province, to be reunited—by force if necessary—with the Communist motherland; and intelligence gathering over international waters, which the Chinese have declared to be off limits to unauthorized access.

Meanwhile, the Americans were particularly anxious to have the maximum information on new Chinese weapons systems, which accounted for the presence of an aircraft packed with computerized electronic equipment, radios, and direction finding systems. These included search receivers to locate active radar and signals frequencies, and long-range radar to track ships, aircraft, and submarines, with short-range radar to intercept communications. All the complex systems were managed by a team of eight cryptographers, with a backup of six more, together with a six-strong navigation and electronics crew, three pilots, and the aircraft commander. Having such an aircraft packed with leading-edge intelligence gathering equipment and specialists land at one of their military bases must have made the Chinese more than happy.

What actually happened after two Chinese fighters intercepted the EP-3 is uncertain. According to the Chinese, the American plane banked sharply to the left, colliding with the tail of one of the fighters and tearing off its tailfin. The fighter crashed into the sea, its pilot ejecting

successfully, although he was never picked up by the rescue services. Damage to the American aircraft forced it to declare an emergency and enter Chinese airspace for an emergency touchdown at their nearest military airfield.

According to the Americans, the Chinese pilot involved, Wang Wei, had approached dangerously close to earlier American flights, even holding up a card bearing his e-mail address when taking up formation with the American aircraft. In this case, the EP-3 was flying straight and level on autopilot, when the Chinese fighter flew underneath it, then pulled up in front of it to force it to alter course, possibly to take it into Chinese airspace, where it could have been forced down legitimately. Whatever the intention, the effect was that the two planes collided, the fighter crashed into the sea, and the EP-3 suffered damage to its port wing, the nose cone, and two of its four propellers. The American plane plunged 8,000 feet before the pilots were able to regain control, and from then on an emergency landing became inevitable.

Realizing that the Chinese would insist on boarding the plane, the crew made maximum use of the time available to them before landing to destroy as much as possible of the secret equipment on board. They used computer self-destruct programs, backed up by powerful magnets, to erase data, and axes to smash the hardware.

Although subsequent negotiations resulted in the crew being returned to the USA, following a tense period of waiting, two questions remain to perplex US intelligence sources. How much information can the Chinese derive from the equipment on board, on US intelligence gathering techniques and capabilities, and the kind of secret information that the United States would be able to gain from these flights? And perhaps even more importantly, what can the future be for further intelligence gathering flights, here and elsewhere in the world's most sensitive political areas, after this incident?

Spying on the Terrorists

ABOVE A fighter for the Taliban Islamic regime in Afghanistan, regarded as a major harbor for terrorist groups, prepares to load shells into a tank turret near the front line against the Northern Alliance troops.

OPPOSITE Freed hostages rescued by German special forces following intelligence information from a hijacked aircraft diverted to Mogadishu.

A s the Cold War recedes into history, one major and most difficult priority remains for the Western world's intelligence and espionage services. The continuing threat of international terrorism maintains a leading place on the list of potential destabilizing influences on the territory of major powers all over the world.

The aftermath of the horrific events in New York and Washington on September 11, 2001, when hijacked airliners loaded with civilian passengers were deliberately crashed into the twin towers of the World Trade Center and the Pentagon, created a new and daunting requirement for Western intelligence agencies.

At one time spies had been used to great effect to contain, and ultimately frustrate the efforts of terror networks in both Europe and the Middle East. The 1970s had seen a huge increase in the use of terror as a political weapon by all kinds of pressure groups, from the IRA in Northern Ireland to ETA in the Basque region of Northern Spain, from the mainly middle-class European revolutionary groups like the Red Army Faction in Germany and the Red Brigades in Italy to the middle-eastern hijackers and bombers like Hezbollah and Black September.

In all these cases, intelligence was vitally important to deter the terrorists. Tracking down terror networks was the exclusive responsibility of the police in the country concerned. In principle though, the law was invariably limited to dealing with terrorist attacks after they had been committed. Preventing future assaults was often an indirect benefit of catching and incarcerating terrorists, though in many cases keeping terrorist leaders in prison often triggers additional attacks to try to force their release.

In other areas, however, fighting the terrorists has become a heavy intelligence responsibility. In penetrating to the heart of these inherently dangerous and secretive organizations, spies and particularly double agents offer the only hope of finding out their plans, and setting up counter-measures in time. Sometimes these operations involve acting on the territory of other countries, but increasingly, as global terror networks extend their reach into the territories of their targets, spies are now operating inside their own countries, but within the alien surroundings of an immigrant religious or political faction.

Because the price of discovery for any undercover agent is instant and final disappearance, few details ever emerge of how and why the different intelligence operations were mounted. Even a review of a successful operation long after its completion can implicate and endanger those involved, even when names and other details remain secret.

ABOVE Oberst Ulrich Wegener, commander of the German Special Forces unit GSG9 is decorated by President Scheel for his role in liberating the Mogadishu hostages.

In the circumstances, all that can usually be done is to make informed conclusions from the little information released by official intelligence sources, as to the kind of information which made it possible to prevent atrocities, release hostages and seize the planners and foot-soldiers responsible. Most successful terrorist organizations have learned the value of the security principles established by intelligence services and the wartime resistance movements.

Penetrating the terrorists' security

By dividing large groups into small cells and keeping contact between adjacent cells to a minimum, terrorist groups can reduce the damage caused by an individual being identified or captured. By taking the utmost

ABOVE RIGHT Aldo Moro, Foreign Minister of Italy and later Premier during talks at Downing Street in London.

BACKGROUND Police with crowds of onlookers at the site of the Moro kidnapping.

BELOW The body of former Democratic Party President Aldo Moro left in a car in the center of Rome after his kidnap and murder by the Italian Red Brigades.

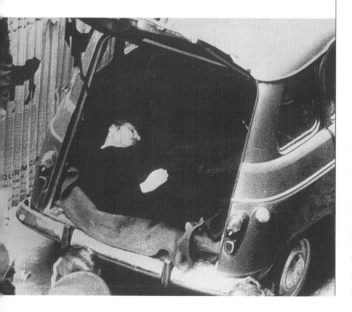

care over recruiting new people, they can limit the chances of the counter-terrorist services placing a double agent in their midst and by exerting swift and terrible retribution against any of their personnel suspected of betrayal, they can keep the tightest of grips on those working for them.

This presents the counter-terrorist groups with great difficulties. In most cases, they can only hope to place someone on the inside of a terrorist group by recruiting volunteers from within the same nationality, population, religious or political group as the terrorists themselves. In trying to recruit volunteers from within such a restricted group, they leave themselves open to the old espionage weakness of a double agent turning out to be a triple agent—the person they recruit to relay information from within a terrorist organization may be working with the knowledge and support of the terrorists, to betray those with whom he or she comes into contact, and leading to counter-terrorist information being passed to those against whom it is targeted.

In other cases, where the agents are genuinely on the side of the counter-terrorist service which runs them, they put themselves into great danger even by the simple act of entering the community from which they might make contact with the terrorists. Should they manage to be accepted and recruited, they may find they only come into contact with other individuals in their own cell, at a low level in operational terms. To try to find out details of policy, targets or forthcoming operations, they have to make themselves conspicuous, and thereby move into positions of still greater jeopardy.

However, intelligence services have been able to prevent a number of terrorist attacks through timely information, aided by progressively higher levels of security. Perhaps as a direct result of these successes, there has been a tendency for fewer terrorist attacks, but with a dramatic increase in the average death toll. Particularly among the Middle Eastern terror networks, suicide and car bomb attacks are the preferred method, given an apparently limitless supply of dedicated agents who accept death in making a successful attack. These included the truck bombs against the US

embassies in Nairobi and Dar-es-Salaam in August 1998, and the suicide attack against the *USS Cole* in Aden harbour in October 2000.

For the time being, the security services are still at a loss as to the best way to counter suicide attacks. So far, the main response has been purely defensive. When Osama bin Laden's agents attacked the American embassies in East Africa, the US reaction was to close down most embassies world-wide until it could be seen whether or not more attacks were on the agenda. Assassinations too have been difficult to counter, no matter how well motivated and hard-working the security services, when those responsible are often all too happy to achieve martyrdom along with

BELOW Some 80 people were killed and more than 1,000 injured in the two large car bomb attacks on the US embassies in Kenya and Tanzania in 1998.

ABOVE A body is removed from the wreckage of the US embassy in Nairobi, Kenya after the 1998 car-bomb explosion.

their victims. These have included President Sadat of Egypt, hated by fanatics for his role in helping to improve the prospects for peace in the Middle East, Prime Minister Rabin in Israel for similar motives from the other side of the Arab-Israeli divide, and Prime Ministers Indira and Rajiv Gandhi in India.

BELOW Volunteers assist with the mammoth task of rescuing victims of the Nairobi bomb.

Counter-terrorism successes

There have been counter-terrorism successes, however. These include the American campaign to track down the agents behind the first attempt to blow up the World Trade Center towers in 1993. Like the recent and infinitely more destructive attack, this involved the help of a number of terrorist organizations, including the likely involvement of bin Laden, to carry out a blow against the symbol of what they saw as the American-dominated global trade system, principally as a punishment for US support of Israel, but also as an outlet for deeper and darker feelings of anti-Americanism.

In terms of its objectives, this could have produced a death toll even higher than that of September 11. The plot involved placing a large bomb in the underground car park at the base of one of the towers. With luck, the plotters hoped, this would be enough to cause the tower to collapse against the second tower, bringing both down across the heart of the New York financial district causing tens or even hundreds of thousands of deaths. Fortunately for those who would have been its victims, however, the raiders were astonishingly inept, though the planning had been so careful that the intelligence services had little or no advance warning.

On February 23, 1993, a yellow van was parked in the garage below the North Tower of the World Trade Center. It contained a massive charge of some 1200 pounds of explosive which, when detonated, killed six people and injured more than a thousand, causing more than 500 million dollars damage. A massive FBI investigation was mounted to catch those responsible, and clues soon began to emerge.

BELOW Two undated file photographs of Islamic militant Ramzi Ahmed Yousef who was accused of planting the truck bomb detonated below the World Trade Center in 1993, which killed 6 people and injured 1,000 others.

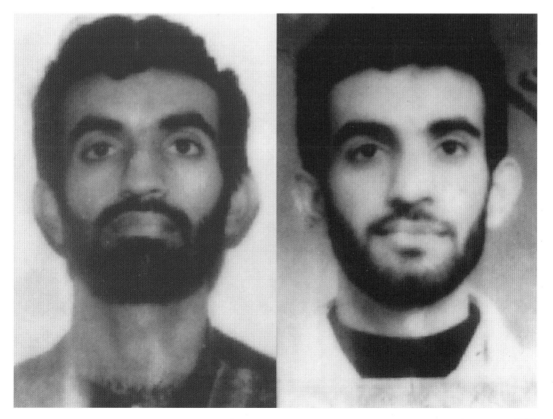

IRA and ETA:

One of the longest-established urban terrorist movements in Europe is the Irish Revolutionary Army (IRA), first created in January 1919 to force the UK government to grant Ireland total independence. After a violent two-year campaign of sabotage and raids by groups of armed men, murdering the British agents who attempted to monitor their operations, the British granted Ireland independence as a Free State, with Dominion status within the Commonwealth.

Disputes over this 1921 settlement caused the IRA to split between those who opposed the political settlement as an unacceptable compromise with their original intentions, and those who accepted it. The opponents took up arms once again, and the resulting Civil War cost more Irish lives than the original campaign against the British.

The Irregulars, as the opponents of the settlement had become known, were finally forced to surrender. The organization hid their weapons and went underground, to re-emerge when the British became involved in the fighting of World War II. The IRA mounted attacks inside Britain to disrupt the war effort, and the Irish government declared the organization illegal, imprisoning many of its agents and executing five of its leaders.

By the late 1960s, the dormant IRA found a new cause. Under the original Irish settlement, the Ulster Protestants had never accepted rule from Dublin, and formed their own armed resistance movement. Their refusal to join an independent Ireland led to partition—the Free State included the 26 counties in the south, east and west of the island, leaving the six counties of Ulster in the north to continue as part of the UK.

Unfortunately, Ulster contained a substantial population of Roman Catholics, whose sympathies were with their compatriots and fellow Catholics in the south. Relations between the two communities worsened until in 1968, the Catholic community in Ulster began a civil rights campaign to redress discrimination over jobs and housing. When this provoked a violent reaction from Protestant extremists, the IRA entered the fighting with a self-proclaimed mission to save the Ulster Catholics, but with a hidden agenda to bring about the original 32-county all-Ireland state as a socialist republic.

Once again, the IRA split over the policy of violence. The Marxist faction became the Officials, and gave up using violence in 1972. The remaining Provisionals maintained a terrorism campaign in both Northern Ireland and on the British mainland, causing the deaths

of more than 3,000 people over almost three decades, until the series of cease-fires and the development of the Peace Process during the 1990s.

This long campaign never achieved its objectives, despite the suffering of Ulster people, and the casualties among the army, the police and innocent civilians. The IRA insists the majority of people of Ireland as a whole want reunification, with Ulster being taken from the UK and absorbed within the Irish Republic. The Unionists insist that within Ulster itself, the majority is equally determined not to be any part of a united Ireland, no matter how long or bloody the violence of the campaign to force them to do that.

During this long struggle, intelligence forces attached to the Royal Ulster Constabulary and the British Army have faced crucial advantages and limitations because of the nature of the theater of operations and the terrorists against which they and their agents have had to operate. Furthermore, in spite of unique rights and freedoms accorded to citizens of the Republic of Ireland living within the UK, anti-terrorist operations on the British mainland have managed to prevent a whole series of planned atrocities, in spite of the occasional high-profile disaster like the bombing of the Conservative party conference in Brighton, and the mortar attack on the Prime Minister's official residence at 10 Downing Street.

For the present, terrorism in Northern Ireland is on hold, though as yet the IRA has not put so much as a single bullet beyond use. The other parties to the agreement vary between anger at negotiating with organizations who remain armed to the teeth and able to return to violence whenever they feel they might not achieve their intended purpose, to those who feel almost any price is worth paying, if terrorist arms remain in storage rather than in use.

In many ways, the fight maintained by the Spanish authorities against the terrorist organization ETA parallels that between the British and the IRA. ETA's initials stand for "Basque Homeland and Liberty" in the Basque language, and the organization was formed in 1959 when younger and more radical activists split away from the Basque Nationalist Party, which had survived as a clandestine organization during the years of the Franco dictatorship. As with the IRA, the ETA organization split into two factions during the 1960s, the moderates who simply wanted more independence for the Basque region of north-eastern Spain, and the radicals who wanted to begin a campaign of violence to establish a Communist Basque homeland.

While Franco was still in charge, Spanish reaction was equally tough, with ETA suspects being arrested, tortured and imprisoned without trial, until most of the movement's leaders had been captured by the early 1970s. After Franco's death in 1975, the newly democratic Spanish Government gave the Basque territory more regional independence, and offered an amnesty to ETA members who turned their backs on violence. Their reward was a campaign of bombings and murders which reached a level of ten times those it had achieved against the Franco regime.

The greatest difference between ETA and the IRA is that the Basque group only represents a minority opinion both within Spain in general and in the Basque region in particular, so its continued campaign seems unlikely to achieve the results it wants. However, the Spanish authorities' intelligence efforts to track down its leaders are complicated by the fact that the Basque political organizations were forced to run their operations from the adjacent areas of France, where a substantial Basque population lives under French rule with no apparent discontent, during the Franco years. Consequently, many of their offices and leaders remain out of Spanish reach in the territory of another country, though increasing efforts to establish a pan-European anti-terrorist intelligence organization may make tracking them down a more practical possibility.

With the IRA, on the other hand, the British Army has been able to mount a number of highly effective intelligence operations against its terrorist adversaries. Many soldiers in Army units come from the same communities which shelter and support the terrorists, and have the background knowledge of culture and language to merge into those communities to keep a close watch. Some

ABOVE The scene of a car bomb explosion, the fourth in a week in February 1997, in the Basque city of Bilbao. A policeman was killed.

OPPOSITE June 1996. Two hundred people are injured in a truck bomb blast, believed to be IRA, in the center of Manchester, England.

BACKGROUND February 1996. The remains of a London bus after a suspected IRA explosion ripped through it, killing one person and injuring eight others.

have been recruited by terrorist cells for the lonely and dangerous life of the double agent, while others have taken part in the scarcely less lonely and dangerous role of controlling double agents.

In some cases, as with Captain Robert Nairac, the dangers of moving into Republican areas to communicate with agents led to their kidnapping, torture and death. Even serving soldiers watching known Republicans have to be careful not to be in the wrong place at the wrong time. Two Royal Signals soldiers took a wrong turning and found themselves confronted by an IRA funeral procession. Before they could retrieve the situation, the crowd dragged them from their car and beat them to death.

With scarcely credible stupidity, one of the conspirators, Mohammad Salameh, called at the Ryder rental office to claim the $400 deposit he had paid when the van was rented, claiming it had been stolen from him by Ramizi Ahmed Yousef, one of the bomb-making team, the day before the blast. FBI agents traced the leads back to a Jersey City apartment and a nearby storage shed used as a bomb-making factory where they found the fingerprints of the two men and another conspirator, Eyad Izmoil.

Both Yousuf and Izmoil had left America on a flight from Kennedy airport to the Middle East immediately after the explosion. But the long arm of the intelligence agencies pursued them across the world, aided by a $2 million reward for information leading to their capture. Yousef was reported in the Philippines and in Thailand and was finally arrested in Pakistan, while his accomplice was captured in Jordan. Almost five years after the bomb went off, they and four conspirators were tried in the US and sentenced to life imprisonment.

Another success of the fight against terrorism resulted in the tracing and conviction of Timothy McVeigh, the Oklahoma bomber. A former soldier who had served in the Gulf War, McVeigh placed an even larger bomb aboard another Ryder rental truck which he drove into the basement car park of the Alfred P Murrah Federal Building in Oklahoma City on the April 19, 1995. The resulting explosion killed a total of 168 people, many of them young children attending a day-care center in the building.

Once again, the agents investigating the crime had a liberal ration of good luck to reinforce their hard work. Among the vast quantity of rubble which had to be sifted for significant fragments, they found part of the truck axle which revealed a partial vehicle identification number. This was traced to a 1993 Ford truck which had been hired from the Ryder Rentals office in Junction City, Kansas. Staff at the office helped artists reconstruct computer-generated pictures of the two men who had rented the truck, and these images were issued to more than 1000 agents, who began questioning staff at gas stations, hotels and restaurants between Junction City and Oklahoma City.

At last, the patient checking bore fruit. At the Dreamland Hotel in Junction City, the manager remembered a guest who looked like one of the pictures and who had been driving a Ryder truck. The register revealed the name Timothy McVeigh. Even this luck was eclipsed by the agents' next stroke of good fortune. When they tried to determine McVeigh's whereabouts, they found he was already in police custody, having been arrested for carrying a semiautomatic pistol, and was being held at Noble County jail. Forensic evidence showed explosives traces on the suspect's clothes, and empty barrels at the Michigan address on his driver's licence were very similar to fragments found at the bomb site.

McVeigh was eventually executed, though he never identified any other conspirators, or any group for whom he was acting.

More recently, there has been a perceptible shift in the basis for Islamic terror groups. At one time these were principally aided and supported by Iran to operate among the Shi'a Muslims. Since then, terror groups have been created among their hugely more numerous Sunni equivalents, who represent the overwhelming majority of Muslims. Recent additions to terror within Muslim countries include Algeria, Indonesia, Pakistan, Saudi Arabia and Central Asia. Also intelligence reports are said to have revealed a plot by Afghan immigrants in New Zealand to attack a

BACKGROUND Timothy McVeigh was arrested and charged for the Oklahoma City bombing, which killed 168 people. He was found guilty, sentenced to death and later executed for his crime.

ABOVE A handcuffed Timothy McVeigh is escorted from the Noble County courthouse after originally being taken into custody for carrying an illegal semiautomatic pistol.

nuclear reactor near Sydney at the time of the 2000 Olympics. Currently several nations are thought to be sponsoring terrorism, including Sudan, Syria, Lebanon, Iran (though this is now officially denied) and Afghanistan.

Though much of the terrorist threat remains rooted within factions of the Islamic community, there are other powerful organizations outside the Muslim world, some with rational political philosophies, others without.

Nerve gas on the Tokyo underground

One of the most terrible events, before September 11, was the sarin gas attack by the religious zealots of Aum Shinrikyo in the Tokyo underground in March 1995.

The founder of the cult, Shoko Asahara, predicted that world war would break out with America delivering gas attacks on Japanese cities before switching to nuclear weapons. Japanese agents monitoring the cult's activities found documents which implied the cult was willing to provoke this catastrophe with a pre-emptive strike in 1995. In April 1994, Asahara claimed Japanese and US military aircraft had already bombed the cult with sarin gas. Two months later, on June 27, 1994, sarin gas was released in the town of Matsumoto, resulting in the deaths of seven people. Suspicion fell on the Aum Shinrikyo, but in spite of their earlier statements, the cult strongly denied responsibility.

Consequently, the sarin attack on the March 20, 1995 came as a profound shock to the authorities. In all, 12 people died and more than 5,500 were injured, a toll the terrorists undoubtedly found disappointing. Though they used teams of trained technicians in a massive operation, they killed fewer people than would have died from the effects of a conventional bomb, probably because of impurities in the normally highly lethal gas.

A later attempt to launch an anthrax attack in Tokyo failed completely, though Asahara and 40 followers did visit Zaire in 1992 with the humanitarian aim of helping victims of the deadly Ebola virus epidemic raging at that time. With 90% mortality and no known antidote or treatment, this fearsome disease makes an ideal terror weapon, and official sources in the USA believe the real purpose of the 1992 visit was to obtain samples of the virus to be cultured for later use.

Even within the USA, there are groups contemplating similar outrages against society. An Ohio laboratory technician named Larry Harris ordered bubonic plague bacteria from a Maryland supplier of biomedical supplies just weeks after the Tokyo incident. When company representatives became suspicious, they contacted the authorities, who found Harris was a member of a white supremacist

Iraqi terrorism—believe that today's terrorists lack the know-how and the technology to produce and deliver these agents on a large enough scale to cause the casualties they seek.

Terrorism on a global scale

In more recent times, terrorist organizations have used the large-scale movements of asylum seekers, and the freedom and openness of Western society to further their aims. Among the many migrants to America and Europe from nations like Egypt, Iraq, Iran, Pakistan or Afghanistan are a small proportion of agents from the different terror networks. For the time being, their operations in the West have been largely confined to using their host countries as bases for recruitment, supply and funding for their operations within their target nations.

Fortunately for the intelligence services, the increasing globalization of the terror networks

ABOVE Members of a specialist anti-gas chemical warfare unit of the Japanese Army check out a Tokyo subway station following the 1995 sarin gas attack.

RIGHT Japanese army soldiers with gas protection gear prepare to enter the Kasumigaseki subway station after the sarin gas attack.

organization. When questioned, he claimed to be using the bacteria to develop a defense against the introduction of Iraqi super-rats carrying plague germs to America.

Chechen terrorists were said to have used low-grade nuclear materials in their attacks in the forests surrounding Moscow in March 1995, and Saddam Hussein's regime in Iraq has used biological and chemical agents, even against his own people. However, many experts—including the Israelis, always a prime target for

opens up new potential fault-lines which allows them to be uncovered in different ways. To begin with, a network in a Western society, actively looking for new recruits, is going to be easier for agents of the right background and culture to penetrate than one hiding in the mountains of Afghanistan or the deserts of Iran, where any approaching stranger is going to be under the deepest suspicion.

Secondly, and even more importantly, these more widely dispersed terror networks create communications problems for those directing them. Much of their publicity and recruitment information is now broadcast over the Internet, which enables intelligence agencies to understand the motivation of their adversaries, however warped, and enables them to determine links and structures of these shadowy organizations. The use of spy satellites to eavesdrop on microwave links and telephone conversations, in spite of the prodigious volume of messages which can be

ABOVE A soldier of the Afghan Taliban movement firing artillery north of Kabul.

overheard, has revealed gems of information, like Osama bin Laden's instructions to his networks to go to ground in Afghanistan before the September 11 operation.

Obstacles to counter-terrorism

How can intelligence services hope to fight these threats at home as well as in the most remote parts of the world? Penetrating these highly motivated, ultra-religious groups calls for agents with formidable specialist expertise as well as extreme courage and motivation. Potential agents need training in the cultural background of the groups they plan to infiltrate, and this includes complete fluency in the colloquial speech and dialects of minority languages like Pushtu or Farsi, in addition to mainstream tongues like Arabic or Urdu.

207

Yet the CIA in particular has faced two almost intractable problems in placing agents into the terrorist networks posing the most direct threat to the West. Former CIA operative Reuel Marc Gerecht revealed a dearth of officers within the Agency with the ability to speak fluent Arabic or the appearance and background to pose as a member of bin Laden's terror network. Even where the necessary knowledge and people existed, the motivation to spend months or years in the austere environment of the Afghan mountains was lacking, compared with targets in the more congenial and less remote terrorist centers like the West Bank and the Gaza Strip.

The other problem originated outside the Agency, and among its political masters. A report from the National Commission on Terrorism, presented to Congress in July 2001, revealed that the CIA had been discouraged from recruiting undercover sources of information within terrorist networks because of Government regulations put in place in 1995. These allows intelligence officers to be held personally responsible for any illegal acts or human rights violations carried out by agents they had recruited. In addition, the Justice Department has tended to refuse permission for telephone tapping to be carried out on suspected terror groups because of possible damage to their human rights, and even where permission was granted, there was a lack of enough expert Arabic speakers to monitor and translate the hours and hours of conversations recorded as a result.

Another limitation on CIA action was the ban on political assassinations, introduced by the Ford administration in 1976, as a result of pressure by the House Intelligence Committee. Previously, the Agency was said to have killed as many as 20,000 members of the Viet Cong during the Vietnam campaign, and to have mounted as many as 26 unsuccessful attempts on the life of Cuban President Fidel Castro. The Clinton administration then introduced a blanket ban on recruiting any agents or even

ABOVE Aerial view of Ground Zero—the base of the World Trade Center twin towers, showing the destruction of the towers and adjacent buildings.

informers with criminal records or who had been guilty of human rights violations. Nevertheless, details have emerged of an attempt to assassinate Osama bin Laden, which had originally been backed by President

working alongside the intelligence services of other countries with a closer geographical or cultural relationship to the nations who harbour them. In the case of terror networks based in Afghanistan, for example, the Pakistani Inter Services Intelligence Agency (ISI) is based on Khayban e Suhawardy Avenue in Islamabad, where some 100 officers run a large network of agents operating within Afghanistan.

In the meantime, there have been signs of a relaxation on the official restrictions which so handicapped the efforts of the CIA to monitor terrorists all over the world. Since 1998, and the attacks on the US embassies in Africa by bin Laden's network, presidential directives imposed new rules of engagement on CIA agents in the field. Though the ban on attempting to assassinate a foreign head of state remained, agents were allowed to use lethal force in their own self-defence, which in practice meant terrorists could be killed immediately before they mounted an attack.

Though the Agency refuses to comment, other sources claim that CIA men, in collaboration with foreign intelligence agencies have been able to prevent attacks by bin Laden's al-Qaeda network in Kenya, Egypt, Jordan and in Europe. Only time will tell whether the spies and intelligence agents have enabled the whereabouts of the terrorist chiefs to be pin-pointed accurately enough for them to be hit by the military attacks designed to eliminate the threat they present. Meanwhile, within the United States, the authorities are calling for the development of more effective defences against the terrorists, including new sensors to detect nuclear weapons in transit (for example, gamma-ray imaging systems, including stimulation to elicit detectable emissions), new types of 'tripwires' suitable for many different entry-points, like explosives-sniffers and body scanners, and their prototyping for mass-production, and development of advanced anti-viral agents to be used against the threat of smallpox.

Clinton in 1998, but the project was overtaken by political events in Pakistan.

Fighting tomorrow's terrorist

It may well be that Western intelligence agencies have to reinforce their attempts to place their own agents into the terror networks,

European Terror

ABOVE Former Italian premier Aldo Moro speaking at a press conference in the Quirinale Palace in Rome, before his abduction and murder by Red Brigades terrorists in 1978.

BELOW Entebbe airport, Uganda, where Israeli commandos mounted the 1976 raid to liberate hostages from an aircraft hijacked by the infamous Baader-Meinhof gang.

During the 1970s, terrorism was rife in Europe with the largely middle-class movements like the Red Brigades of Italy, and the Red Army Faction in Germany. The Italian movement began as the creation of a left-wing former student named Renato Curcio who formed a radical group at the University of Trento in 1967 which devoted their time to studying the teachings of Marx, Mao and Che Guevara. He married a fellow radical two years later and moved to Milan where they announced the formation of the Red Brigades in 1970. The movement shifted into more violent behavior, beginning with firebombings of commercial premises, then turning to kidnapping to win notoriety and raise funds and finally to murder of carefully selected targets.

The group grew more ambitious as well as more ingenious. In 1974 they kidnapped and then murdered the chief inspector who was head of the anti-terrorist unit in the Turin police force. Four years later they kidnapped the former Italian Prime Minister, Aldo Moro. They threatened to murder him if 13 of their number, then languishing in jail, were not immediately released. When the authorities refused, Moro was killed.

Gradually, through careful intelligence work, the authorities began to redress the balance. In 1981, the Red Brigades kidnapped a senior American officer serving with NATO, Brigadier General James Dozier. After six weeks in captivity, he was liberated when Italian police raided the group's safe house in Padua. With more and more of their agents, and later their leaders, arrested and jailed, the influence of the organization, which numbered several thousands in its mid-1970s heyday, went into terminal decline.

A similar bloodthirsty ideology governed the operations of the German Red Army Faction, which was also known as the Baader-Meinhof group, after Andreas Baader and Ulrike Meinhof, two of its founding figures. These too specialized in robberies, burnings, kidnappings and murders, with the added dimension of links to Palestinian terror organizations. When the Palestinians hijacked an Air France airliner in 1976, two members of the Red Army Faction formed part of their team. The flight was bound from Israel to France, and after a stop at Athens, the hijackers took over and forced the crew to divert to Entebbe in Uganda. There most of the passengers were freed, apart from those who appeared Jewish, who were held hostage for the release of 53 Palestinian prisoners.

In an intelligence coup masterminded by the Israeli Mossad organization, enough information was amassed to enable a force of Israeli commandos to mount a raid on Entebbe. Carried the 2,500 miles to the target aboard four C-130 transports escorted by Phantom fighters, the Israelis killed seven of the terrorists and liberated all but three of the hostages, and lost one soldier in the operation. By that time, both Baader and Meinhof had been tracked down, arrested and imprisoned, along with most of the key members of the group. Meinhof committed suicide in that same year, but the others pinned their hopes of release on the planned hijacking of a Lufthansa airliner to Mogadishu in Somalia, where passengers and crew were to be held hostage for the release of the imprisoned terrorists. This too ended in a victory for the counter-terrorism forces. West German commandos stormed the plane and released the hostages on October 17, 1977. On the following day, Baader and two other terrorists were found dead in their cells. The final blow to the group's viability came with the collapse of the East German regime in 1990, when the files of the Stasi secret police revealed that they had helped train, support and protect the terrorists as a means of attacking the West during the Cold War.

ABOVE Relatives wait in anguish for news of the 104 hostages held for more than a week by Palestinian hijackers in Entebbe.

OPPOSITE The remains of Dora Bloch, an Israeli grandmother who was killed in the Entebbe hijacking, are officially handed over to her son by the Ugandan authorities three years later.

The Lockerbie disaster

Prior to the 2001 attack on the World Trade Center, one of the most spectacular and horrific terrorist atrocities against civilian victims took place at three minutes past 7 on the December 21 1988. A Pan American Boeing 747 on a flight from London Heathrow to New York with 243 passengers was crossing over south-western Scotland at its cruising height of 31,000 feet, when air traffic control radar operators noticed its echo had broken up into several smaller traces, which were rapidly fading from the displays

A bomb carried in a suitcase in the hold of the aircraft had detonated, splitting the airliner's outer skin. This was torn apart in the fierce slipstream, so weakening the aircraft structure that the huge 747 broke up. The nose section was torn away, and the collapsing airliner fell upon the small Scottish town of Lockerbie, killing all on board, and destroying 21 houses with another 11 victims on the ground.

To find the cause of the accident, investigators traced more than four million pieces of wreckage collected over almost a thousand square miles of land. These then had to be reassembled piece by piece to reconstruct nine-tenths of the aircraft, so that the order of its break-up could be explained and understood.

The experts found tiny bomb fragments, with traces of Semtex explosive and a piece of printed circuit board identified as part of a Toshiba radio-tape-player, trapped in the skin of the luggage container at the site of the fatal explosion. Clothing fibers and fragments of a brown Samsonite suitcase were also found, and further tests showed the case had been placed

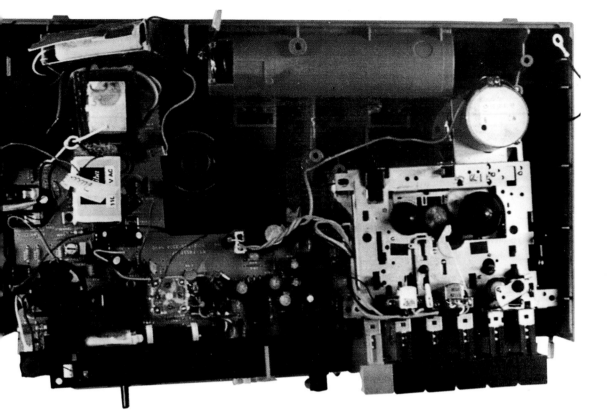

aboard the 747 after being transferred from a connecting flight from Frankfurt.

Piece by piece the intelligence agencies assembled their evidence in turn. They knew that terrorists were aware that security checks at Frankfurt at that time were less strict than in London, and transfer bags which had been checked at Frankfurt would not be checked again when transferred to the New York flight. Furthermore, at that time it was possible for a passenger to accompany bags on the Frankfurt-London sector and simply not board the aircraft for the New York flight as baggage was not then positively matched at Frankfurt against passengers actually on board.

There had also been intelligence reports that a major terrorist action could be expected during this period. Almost six months earlier, on July 3, 1988, the US Navy cruiser *USS Vincennes* was being attacked by Iranian gunboats in the Persian Gulf during the Iran-Iraq war. Following an intelligence warning of increased Iranian attacks around the time of American Independence Day on the July 4, and an earlier attack by an Iranian Mirage fighter bomber more than a year before had killed 37 American sailors

ABOVE A bomb hidden in a cassette player was discovered by West German police when they arrested members of a Palestinian terror group in October 1988, containing 300 grams of Semtex and thought to be almost identical to the one used to blow apart the Pan American 747 over Lockerbie.

OPPOSITE TOP Scottish police guard the special court set up at Soesterberg in Holland, to try those accused of the Lockerbie bombing.

OPPOSITE BOTTOM Police carry the body of one of the Lockerbie victims past the wreckage of the flight-deck of the Pan Am 747.

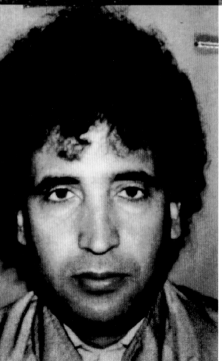

when a missile struck the frigate *USS Stark*, tension was high when an approaching radar echo was misidentified as an Iranian Air Force F14.

The ship fired two anti-aircraft missiles, which destroyed the aircraft, actually a civilian Iran Air A300 Airbus, Flight 655 from Bandar Abbas to Dubai, carrying 278 passengers and a crew of 12, on their way to the important religious festival of Id-al-Adha four days later. Intelligence agents had warned that terrorist groups were tending to place plastic explosives inside electronic equipment in unaccompanied baggage. The CIA had received a specific warning on December 5 that within the next two weeks a Finnish woman would place a bomb aboard a Pan American flight from Frankfurt to the USA. The details were passed to US Embassy staff as a precaution, but investigations by Finnish police found no evidence of anything to back up the story.

Evidence from the site of the explosion revealed clothing fibers, some of which were traced to a Babygro garment which carried a label identifying a clothing manufacturer in Malta, an island with close links with Libya at the time. The trail was followed to a clothing store, where staff remembered a man of Arabic appearance buying a large consignment of clothes, which matched the details of those whose fragments were found at the explosion site. Furthermore, airline baggage records showed that one bag brought in from Frankfurt to be placed aboard the

747 had earlier been sent from Malta to Frankfurt. It was also known to the intelligence agencies that at the time of the attack, Abu Taib, a senior member of the Popular Front for the Liberation of Palestine, was actually in Malta.

The first estimate was that the Iranians had asked the PFLP to mount the attack in revenge for the shooting down of the Iranian airliner in the Gulf in July. However, further investigation traced the Semtex to Libya, and provided a positive identification for the purchaser of the clothes in Malta as one of the Libyans later indicted for the crime. In this case, the intelligence assessment was that the attack was mounted in revenge for an earlier US attack, when bombs were dropped on Libya in retaliation for a Libyan-backed terrorist raid against US forces in Germany. Finally, after intense diplomatic pressure and severe economic sanctions, the two Libyan suspects finally accused of responsibility for the attack were released to stand trial at a special Scottish court set up in Holland, where one of them was finally convicted, almost 13 years after the incident.

Osama bin Laden

What kind of man could find the motivation, let alone the recruits and the organization to mount an atrocity where a whole team of agents died along with their victims, crashing two airliners crammed with passengers and loaded with fuel for a long-haul flight, into the towers of one of the world's tallest buildings? Osama bin Laden is the 17th son of a builder originating from the Yemen in the south of the Arabian peninsula, who made a fortune in Saudi Arabia from a series of building contracts. Bin Laden himself developed his militant views when fighting as a volunteer in Afghanistan, against the invading Soviet forces, where his units were actually aided and equipped by the CIA.

Bin Laden switched targets in 1991, when he found American forces, based in Saudi Arabia for the Gulf War to expel Iraq from Kuwait, were to stay within Saudi borders after the conflict was over to train the Saudi Air Force and the local police, and to provide an additional deterrent to further Iraqi adventures. Citing one particular quotation from the Koran, to

the effect that there should not be two religions in Arabia, he decided the Americans' real purpose was a crusade against Islam, and he blamed their presence for problems with the Saudi economy.

With the passing of time, his views and his actions grew more extreme. In 1996, he declared a "jihad" or holy war against Americans in Saudi Arabia. Bombs in American barracks were followed by attacks further afield in Somalia and, most bloodily, in East Africa.

For this he found much support, both at home and abroad. For example, the al-Khifar Refugee Center in Brooklyn hid behind the façade of a charity designed to help women and orphan children, but was found to have played an active part in the planning of the attempt to blow up the WTC in 1993. It even had a sub-title "Offices for the Services of Holy Warriors" with a headquarters in Pakistan and branch offices in no less than 38 US cities. When it was closed down by the American authorities most of its personnel escaped and went on to play a vital role in setting up bin Laden's al-Qaeda network.

In the aftermath of the WTC atrocity, there were intelligence reports that a potential ally of Osama bin Laden was Imad Mugniyeh, the Lebanese born leader of Hezbollah, sponsored by the Iranians and the man behind the truck bombs which killed more than 300 American marines and French soldiers in Beirut in 1983 and the taking of Western hostages in the Lebanon in the years that followed. The CIA tried to seize him in Saudi Arabia in the early 1990s when he was travelling between Sudan and Syria, but the coup failed because the Saudis withdrew their support at the last moment. He is believed to be living in Teheran, and to visit Syria, Lebanon, the Sudan and Afghanistan on a regular basis, and there are KGB reports to suggest he has been identified in Chechnya, where the truck bombings of Moscow apartment blocks show signs of his handiwork. Bin Laden's truck bombings of American embassies in East Africa certainly seem to owe much to Mugniyeh's methods.

The attacks in 1998, following a second declaration of war against Americans and their allies, civilians or military, anywhere in the world, involved bin Laden's followers detonating two huge truck bombs outside the American

ABOVE The north World Trade Center tower shows a gaping hole where the first airliner collided with it, as the south tower explodes into a fireball when struck by the second aircraft.

OPPOSITE TOP Islamic fundamentalist and terrorist chief Osama bin Laden leader of the al-Qaeda group held responsible for the attacks on the World Trade Center and the Pentagon in September 2001 in addition to the US embassy attacks in 1998.

OPPOSITE Rescue workers aided by volunteers work amidst dust and falling debris to help the countless victims of the World Trade Center atrocities.

ABOVE Firefighters striving to damp down the flames where another hijacked airliner was crashed into the Pentagon.

BELOW A Taliban soldier watching a Russian-made tank firing at hostile Northern Alliance positions around the strategic Bagram airbase.

embassies in Nairobi in Kenya, and in Dar es Salaam in Tanzania, killing 224 people, including 12 Americans. In October 2000 a suicide bomber in a small craft loaded with explosive rammed the American destroyer *USS Cole* in Aden harbour. Finally, in September 2001, his hatred for America resulted in the destruction of the World Trade Center and the loss of some 5,000 innocent lives.

Bin Laden is an intelligence agent's nightmare. He has mounted disinformation and deception campaigns against the US by using decoy teams and sending false messages over

communications channels monitored by Western intelligence, to mislead them over his real targets. One possible deception in June 2001 resulted in the Pentagon calling off military exercises in Jordan and ordering the US 5th Fleet to sea, with US forces in the Gulf put on the highest state of readiness. To this day, no one knows if the American reaction was a response to an imaginary threat or whether it foiled a genuine operation. Other intelligence services claim that they have uncovered evidence that bin Laden's organization planned to attack the European Parliament with nerve gas.

US intelligence had monitored bin Laden's satellite telephone conversations since 1996, but more recently his al-Qaeda network has switched to sending messages by couriers carrying coded instructions to be relayed from telephones in Pakistan. A rumour that he was visiting the Zhawar Kili al Badr training camp in eastern Afghanistan after the Embassy bombings to meet leaders of other terrorist groups resulted in the US Navy firing 70 cruise missiles at the area, only to find he had left the camp before the strike.

BELOW The exterior ironwork of the World Trade Center tower over the crews carrying out the massive cleanup operation once all hope for trapped victims had faded.

At one time, the Americans believed he was based in a complex of mountain caves near Jalalabad in eastern Afghanistan, but that after the US strikes, he was thought to have moved to a new base in the Pamir Mountains, travelling in darkness in convoys of black all-terrain vehicles, escorted by Arab volunteers. Operating against such a remote and primitive background poses real problems in locating the target, even for services lavishly equipped with satellite surveillance and signal intelligence.

More recently, US intelligence services were authorized by President Clinton in 1998 to mount an attempt to capture or assassinate bin Laden. Squads of American special services commandoes were sent to the remote Parachinar camp in North Pakistan, close to the Afghan border, for training in July 1999. The collaboration involved Pakistan's Prime Minister, Nawaz Sharif, and the Pakistan Inter Services Intelligence Agency, who would provide retired intelligence agents as guides to find bin Laden. The mission was almost ready to depart when Nawaz Sharif was deposed in the coup which brought the present leader, General Musharraf to power, and the project was abandoned.

TOP Fundamentalist Taliban fighters show their loathing of alcohol by crushing beer cans under their tank tracks.

ABOVE Happier days for the Taliban in 1997 as their forces capture a town previously held by the opposition, following the local commanders switching allegiance.

Abwehr the German intelligence service from the 1920's until 1944.

Agent A person, usually a foreign national who has been recruited from an intelligence service to perform clandestine missions.

Asset a clandestine source or method, usually an agent.

Audio surveillance operation a clandestine eavesdropping procedure usually with electronic devices.

Black operations clandestine or covert operations not attributable to the organization involved.

BND Bundesnachrichtendienst; the West German foreign intelligence service established in 1956.

Cheka Russian secret police founded in 1917 to serve the Bolshevik Party; one of the many forerunners of the KGB.

CIA The Central Intelligence Agency of the United States of America formed in 1947 to conduct foreign intelligence collection and counterintelligence operations.

Cipher a code where numbers or letters are systematically substituted for open text.

Clandestine operation an intelligence operation designed to remain secret as long as possible.

Code a system used to obscure a message by use of a cipher or by using a mark, symbol, sound or any innocuous verse or piece of music.

COMINT communications intelligence usually gathered by technical interception and code breaking but also by use of human agents and surreptitious entry.

Concealment device any one of a variety of devices used to secretly store and transport materials relating to an operation.

Controller often used interchangeably with handler but usually means a hostile force is involved, i.e., the agent has come under control of the opposition.

Dead drop A secret location where materials can be left concealed for another party to retrieve. This eliminates the need for real time contact in hostile situations.

Defector a person who has intelligence value that volunteers to work for another intelligence service. They may be requesting asylum or can remain in place.

Double agent an agent who has come under the control of another intelligence service and is being used against his original handlers.

ELINT electronic intelligence usually collected by technical interception such as telemetry from a rocket launch collected at a distance.

Handler a case officer who is responsible for handling an agent in an operation.

Hostile (service, surveillance, etc.) term used to describe the organizations and activities of the enemy "opposition services."

HUMINT human intelligence, collected by human sources, such as agents.

Infiltration (operation) secretly or covertly moving an operative into a target area with the idea that their presence will go undetected for the appropriate amount of time.

KGB all powerful intelligence and security service of the USSR during the Cold War. Successor of Cheka.

MI5 the British domestic counterintelligence service.

MI6 the British foreign intelligence service.

Microdot a photographic reduction of a message, so small it can be hidden in plain sight or buried, for example, under the period at the end of this sentence.

Mole a human penetration into an intelligence service or other highly sensitive organization. Quite often a mole is a defector who agrees to work in place.

Mossad Israel's foreign intelligence service.

NKVD Soviet security and intelligence organization 1934–1946.

Okhrana secret police under the Russian Czars 1881–1917.

One-time pad (OTP) sheets of paper or silk printed with random five number group ciphers to be used to encode and decode messages.

OWVL one-way-voice-link; short-wave radio link used to transmit prerecorded enciphered messages to an operative.

Pattern the behavior and daily routine of an operative that makes his identity unique.

Personal meeting a clandestine meeting between two operatives, always the most desirable but more risky form of communication.

PHOTINT photographic intelligence; renamed IMINT, image intelligence. Usually involving high altitude reconnaissance using spy satellites or aircraft.

Profile all the aspects of an operative's or a target's physical or individual persona.

Safe house a building or similar site considered safe for use by operatives as a base of operations or for a personal meeting.

Secret writing any tradecraft technique employing invisible messages hidden in or on innocuous materials usually sent through the mails to accommodation addresses. Includes invisible inks and microdots.

Security service usually an internal counterintelligence service but some have foreign intelligence gathering responsibility such as the Stasi.

Signals any form of tradecraft using a system of marks, signs or codes for signaling.

SIGINT signals intelligence; the amalgamation of COMINT and ELINT into one unit of intelligence gathering dealing with all electronic data transmissions.

Stasi East German State Security; including internal security and foreign intelligence.

Timed drop a dead drop that will be retrieved if not picked up by the recipient after a set time period.

Tosses (hand, vehicular) a tradecraft techniques for emplacing drops by tossing them while on the move.

Tradecraft the methods developed by intelligence operatives to conduct their operations.

Index

Bibliography

Between Silk and Cyanide Leo Marks, Harper Collins (1998); **Seizing the Enigma** David Kahn, Arrow Books (1996); **Secret Agent—the true story of the Special Operations Executive** David Stafford, BBC Worldwide Limited (2000); **Blind Man's Bluff—the Untold Story of Cold War Submarine Espionage** Sherry Sontag and Christopher Drew with Annette Lawrence Drew, Arrow Books (2000); **Secret Service—the Making of the British Intelligence Community** Professor Christopher Andrew, Heinemann (1985); **Wolfpack—U-boats at War 1939-45** Philip Kaplan and Jack Currie, Aurum Press, London (1997); **The End of the American Century—Hidden Agendas of the Cold War** Jeffrey Robinson, Simon & Schuster in London (1997); **Most Secret War—British Scientific Intelligence 1939-45** Professor R V Jones, Coronet (1979); **Spies in the Sky** John W R Taylor and David Mondey, Ian Allan (1972); **Spycatcher** Peter Wright, Stoddart in Toronto (1987); **Secret Intelligence—The Inside Story of America's Espionage Empire** Ernest Volkman and Blaine Baggett, Doubleday (1989); **The Friends—Britain's Post-war Soviet Intelligence Operations** Nigel West, Weidenfeld and Nicolson (1988); **Battleground Berlin: CIA versus KGB in the Cold War** David E Murphy, Sergei A Kondrashev and George Bailey, Yale University Press (1997); **The Second Oldest Profession** Philip Knightley, Andre Deutsch (1986); **The Secret Offensive** Chapman Pincher, Sidgwick and Jackson (1985); **The Butcher's Embrace—the Philby Conspirators in Washington**, Verne W Newton, Macdonald (1991); **Next Stop Execution—the Autobiography of Oleg Gordievsky**, Macmillan (1995); **Cold Warrior—James Jesus Angleton** Tom Mangold, Simon and Schuster (1991); **Gehlen—Spy of the Century** E H Cookridge, Hodder and Stoughton (1971); **The Atom Bomb Spies** H Montgomery Hyde, Hamish Hamilton (1980); **An Illustrated Guide to Spy Planes and Electronic Warfare Aircraft** Bill Gunston, Salamander Books (1983); **Spy versus Spy—stalking Soviet spies in America** Ronald Kessler, David and Charles (1988); **Women Spies—who they are and how they operate today** J Bernard Hutton, W H Allen (1971); **Venona—the greatest secret of the Cold War** Nigel West, HarperCollins (1999); **Deep Black—the Secrets of Space Espionage** William E Burrows, Bantam Press (1988); **Mayday—Eisenhower, Khrushchev and the U2 Affair** Michael R Beschloss, Faber and Faber (1986); **The French Secret Service** Richard Deacon, Grafton Books (1990); **Masters of War: Sun Tzu, Clausewitz and Jomini** Michael Handel, Frank Cass, London and Portland, Oregon (1992); **The Codebreakers** David Kahn, Sphere (1973); **War Through the Ages** Lynn Montross, Harper Brothers, New York (1960); **Great True Spy Stories** edited Allen Dulles, Robson Books (1984); **A Century of Spies—Intelligence in the Twentieth Century** Jeffrey T Richelson, Oxford University Press (1995); **Reflections on Intelligence** Professor R V Jones, Mandarin Paperbacks (1990); **Spy Counter Spy—an Encyclopedia of Espionage** Vincent and Nan Buranelli, McGraw Hill Inc (1982); **Piercing the Reich** Joseph Persico, Michael Joseph (1979); **Ultra in the Pacific** John Winton, Leo Cooper (1993); **The Guns of August** Barbara Tuchman, Four Square Books (1964); **Battle of Wits—A history of psychology and deception in modern warfare** David Owen, Leo Cooper (1978); Scripts for programmes 1-4 of **Dirty Tricks—a history of deception in warfare**, broadcast BBC Radio 4 and BBC World Service in 1993, David Owen.

Photo Credits